FARMERS MOON

The Evidence for Increasing Crop Yields through Use of Cosmic Rhythms

by Nicholas Kollerstrom,
M.A.Cantab., PhD.

2nd Edition, Revised, 2019

Farmers Moon

Published by
New Alchemy Press
London, 2013

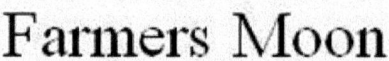

ISBN: 9780-9572-79940

© Nick Kollerstrom, 2013

"The magnitude of the yield deviations that were associated with lunar factors were of practical significance." - H.Spiess, 'Chronobiological Investigations', 1990.

"We have lost the cosmos, by coming out of our responsive connection with it, and this is our chief tragedy. What is our petty little love of nature - Nature! - compared to the ancient magnificent living with the cosmos, and being honored by the cosmos!" - D.H.Lawrence, 'Apocalypse1980' p.76.

"On the days of the Full Moon, something colossal is taking place on Earth." Rudolf Steiner, a lecture on the renewal of agriculture, 7th June 1924.

Dedicated to Frau Maria Thun, pioneer of lunar gardening (1922 - 2012) and to Reg Muntz & Colin Bishop, whose agricultural research work made this book possible.

Contents

Introduction – *page 3.*

Chapter 1 - **Perspectives** *– page 9.*
Kepler's view - Fertility - Sidereal Moon - Organics - Algol.

Chapter 2 - **Four Kinds of Month** *– page 28.*
The Synodic month - The Pull of perigee - The Nodal cycle - The Sidereal cycle - The Tropical Month - A Biune system.

Chapter 3 - **Lunar Phase in Plant Growth-Rhythms** *– page 43.*
Metabolic rate - Water absorption & tree felling - Growth rate - Biochemistry - Bio-electric fields - Plant DNA - Fertilization - Kolisko to Zurcher.

Chapter 4 – **Four Elements in the Zodiac** *- 65.*
The Thun-Heinze sowing trials - Replications - Symbolism - Dr Heinze's opinion - The Spiess Challenge - Chronology: Eight decades of endeavor.

Chapter 5 - **Design of a Time-Experiment** *- 93.*
Considerations of quality - Genesis of the Kimberton Calendar - The Experimenter as subject - Up the garden path - A Rival show - The Concept of significance

Chapter 6 - **Star-Rhythm in Crop Yield** *- 110.*
EARTH: potato & radish yields - FIRE & bean yields - WATER & lettuce yield - Five hundred rows.

Chapter 7 - **The Starry Script** *- 129.*

A Confusion of terms - The Forgotten zodiac - Twelve constellations - The Tropical zodiac of Ptolemy - 20th century boundaries - Paul Platt's version – Goats' milk in the fire-trigon - Tasting the Wine.

Chapter 8 - **Planets and Perennials** *- 148.*

A Traditional view - Rulerships old and new - Improving the wine - Picking time - The Vine, the rose and the apple - Tree bud rhythms .

Chapter 9 - **Horse-Breeding & the Lunar Month** *- 170.*

Timing of Estrus - Mare Fertility - Geomagnetism and sun-node angle - Spring surge in trout-migration - The Public's Interest - GMF.

Chapter 10 - **The Silver Axioms** *- 186.*

View from down Under - Karen's Pear Tree - Cycles of the Sun - The Hour of moonrise - Lunar day - GMF - Equations of the Silver Axioms - Lynx and the lunar node.

Conclusion *- 209.*

Appendices - Bibliography – Index – page 239.

Foreword

Nick Kollerstrom has an established name in Britain for his annual lunar calendar. While in this respect his voice is less well heard that of Maria Thun or Kimberton Hills, it is respected for it's independent approach.

This volume consists of material, some of which has previously been published but, with the benefit of research over the past 20 years, now appears revised and expanded in book form. In truth, it is a lifetime's work and demonstrates wide communication with other workers as well as an extensive presentation of sources.

While identifying itself clearly with the Biodynamic-anthroposophical stream it nevertheless deals in an intellectually-rigorous way with interpretations of astronomical rhythms. Most crucially perhaps, the author's research brings a fresh approach to the use of zodiacs. It is shown that despite the predominant adherence to lunar-zodiacal influences based on the visible constellations, this is not supported by historical precedent or a careful analysis of growth responses. Thus, a system based on the 'sidereal zodiac' – one where the astrological, or tropical, zodiac has been rotated to account for precession – would appear to bring sharper focus to the Biodynamic calendar. This is just one of a number of instances where the book brings new insight into processes accepted but little understood by those working with Biodynamics.

A major contribution is to have provided a comprehensive review of all the evidence from his own and others' research relevant to lunar gardening or farming. This material spans half a century and includes both field and laboratory studies. This too, will help promote a sounder basis for the use and wider acceptance of lunar gardening.

Kollerstrom states that 'the conclusions here presented indicate that organic gardeners or farmers throughout the world could significantly increase their crop yields by taking notice of the Moon's position at sowing time'. From my own experience, including many situations with which I have been involved overseas, I feel confident in supporting this contention. In short, the book should be highly recommended as informing the use of Biodynamic calendars for years to come.

Richard Thornton Smith

Richard Thornton Smith is a former University of Leeds professor and faculty member of Emerson College, Sussex; editor of 'Agriculture', a set of extracts by Rudolf Steiner, published by Sophia Books, Rudolf Steiner Press, and author of 'Cosmos, Earth & Nutrition' 2009.

Introduction

"To everything there is a season, and a time to everything under heaven: a time to be born and a time to die: a time to plant, and a time to pluck up that which is planted..." - *Ecclesiastes, Ch.3:1-2*

'Harvest Moon' by Samuel Palmer 1830

In our time, a new basis is emerging for an area which gardeners have long viewed as mere folklore. Traditions stretching back into antiquity attest that the Moon's position in the heavens at sowing time is relevant for the growth of crops. Modern investigations are finding a new basis for such attitudes, which are helpful for organic farmers and gardeners. The present work aims to compile experimental results and to draw together data published elsewhere over the last two decades. My views and conclusions have developed equally through participating in these years of experiments, and working on farms which use such calendars. I've also investigated data from horse-breeding records.

A dozen or so articles have been published by the author in this general area - in journals horticultural, astrological and cycles research - and this work is an an attempt to restate these, and to draw some of the threads together. This work is intended as a new-millennium synthesis of researches concerning the relation between cosmic cycles and agricultural practice.

The conclusions here presented indicate that organic gardeners and farmers throughout the world could significantly increase their crop yields by taking notice of the Moon's position at sowing time. Such a view, as would have been taken for granted by a Greek or Roman farmer of old, has for long been consigned to the lumber-room of superstition. Instead, farmers now use fertilizers and pesticides to increase yields with little regard to their harmful effects. As organic methods of farming are now becoming an acceptable ecological answer to the problems of pollution and soil exhaustion, it may now be timely to publish this series of studies as a whole.

Let us hope soon to see farmers consulting a desktop computer program at the start of the week, comprising a small cosmic clock or celestial calendar, for advice concerning optimal times - diurnal and monthly, as well as seasonal, for such things as the sowing of seeds, the mating of livestock and the harvesting of herbs. Such a clock would integrate the several factors relevant to give the organic grower choices for the optimal times, and could well be called, 'Farmer's Moon.' The Earth's geo-magnetic field may also here be relevant. This dawning new millennium could well see such a 'cosmo-agriculture' becoming a practical proposition. The present work aims to facilitate this.

One investigates these matters by means of time-experiments, in which crops (in this case) are observed from the point of view of their response to patterns in time. Scientists are trained to assume that the time when an experiment is performed doesn't as such matter, as their experiments are only affected by physical conditions, which in the present case would be factors such as light, temperature, rainfall, seed and soil quality. Relevant as these are, evidence presented in this book indicates that time-cycles of a monthly nature also have a primary significance. Existing lunar gardening calendars are then reviewed from this point of view.

The themes here developed originated within the Biodynamic farming movement, which is a system of organic gardening and farming founded by Rudolf Steiner in 1924.[1] The calendar systems here discussed have come in a large degree from that movement. Biodynamic farmers and gardeners have in this century been pioneers in their aspiration to work with the cosmic pattern in planning their farm and garden schedules.[2] It was while working on such a farm during the year 1970 that I was first faced with the question as to whether evidence could really endorse using a lunar calendar.

---

1 Rudolf Steiner, 'Agriculture', a course of lectures given in 1924, 3rd edition 1974.
2 Maria and Matthias Thun, 'Working with the Stars, A Biodynamic sowing and Planting Calendar', translated and published yearly by Floris Books, Edinburgh (The German title is 'Aussaattage' i.e. 'days for sowing'). Her son Matthias publishes it at their farm in Dexbach. Thun's 'Work on the Land and the Constellations' 1993 describes her approach to the calendar (The German is 'Hinweise aus ser Konstellationsforschung', 'Tips on the Constellation-research'). Her guide to Biodynamic gardening methods, 'Erfahrüngen für den Garten', placing her calendar work in a broader context, is available in English as 'Gardening for Life' (1999 Hawthorne Press).

Introduction

In the early 1980s I with Simon Best co-authored an annual lunar calendar entitled 'Planting by the Moon',[3] followed by annual calendar-wall charts through the 1980s. The book was then re-launched in 1998 by the present writer.[4] This book is a supplement to such a yearly calendar, allowing a fuller discussion of the issues involved, and represents a new phase of the endeavor. In the future, computerized versions of the calendar may make 'perpetual' calendars, which will be intended to stimulate and aid the intuition of the farmer or gardener, not to replace it, and will give options not instructions. Such a program can be localized to a specific region for risings and settings which is hardly feasible for a wall-chart.

The notion that the Moon plays a significant role in plant growth has long remained a 'beyond the fringe' subject, where loss of research position was the expected reward for taking it seriously. After the heady days of the '60s, when America was going to the Moon and papers on the subject were OK, only a faint trickle of reports on the subject continued. Then in the 1990s, two Parisian biochemist showed how cellular DNA in plants varies in composition with the lunar cycle. This was followed by the substantial Swiss forestry work of Zürcher, showing within a fairly practical context how trees were attuned to the lunar month.[5] The silent symphony of lunar rhythms, the ebb and flow of the tides of life, continues in the plant realm, as indeed it does in all living things, and let us hope that it will come more to the centre stage in this new millennium.

The situation improved once I gained access to the racehorse-breeding records. The breeding of racehorses is one thing which the British do take seriously. Britain may slowly be losing its edge as other nations buy off the best stock. In Mediaeval Europe, the Arabs bred the best horses, and their deep knowledge of astrology was surely relevant, by way of selecting the dates and times for breeding. I have no secrets for improving breeding stock, but have discovered something about fertility, as a pattern in time. There is probably no other species for which the date for each covering and birth date is carefully recorded. The results here presented may not be specific to horses, merely that the data is only available for them. The results of my researches, published in the UK's Equine Veterinary Journal[6], did not exactly cause a stir.

---

3 S.Best & N.Kollerstrom, 'Planting by the Moon 1980-82 including Grower's Guide', Foulsham's, then 1983-1984 Astro Computing Services, San Diego, California.
4 N.K. 'Planting by the Moon, a Gardener's Calendar:' Prospect Books 1999, then Foulsham's yearly from 2001. For a discussion of differences between my calendar and that of Thun see Q & A section in Star and Furrow Summer 2008 by Ian Bailey.
5 See Chapter 3; or web-article, 1997, 'Chronobiology of Trees'.
6 N.K. and Camilla Power, 'The Influence of the Lunar Cycle on Fertility on two Thoroughbred Stud Farms' *Equine Veterinary Journal,* January 2000, 32, 75-77. For my two more recent publications on this topic, see Bibliography.

As a sign of the times, in 2001 the BBC adopted a lunar-gardening section on their gardening website, in which I gave daily mundane advice. Some decades ago the topic was merely a joke, whereas it is now becoming far more acceptable. I have a dream, that there exist the 'silver axioms', celestial principles of practical utility, on which the lunar gardening or cosmo-agriculture calendars of futurity will be based. We can find these. A report by Zurcher on his tree-rhythm research was published in an astronomy journal, 'Earth, Moon and Planets' in 2001. This seems a fine omen for the dawn of a new millennium. Let us hope that astronomy departments will soon carry modules on the astronomical periods found in living things.

This opus was conceived as a way of presenting experimental data, yet vast issues and implications kept arising, concerning the nature of life, its adaptation to cycles of time, the four elements and so forth. Why should one moment of time be of especial importance for an organism, when a seed germinates? Supposing it could be shown that seeds tended to germinate optimally at one portion of the lunar month; and that horses, too, tended to be most fertile then; supposing it could also be shown that the female cycle was in some archetypal sense - or perhaps from some ancient historical connection - aligned to that same monthly cycle, what then? These are deep and centrally important questions, and they do belong to this exposition, but on the whole I have restricted such more general questions because the prime focus of this work has to be practical.

Chapter Seven describes crop-yield experiments, in the form of systematic garden trials, designed to test the several theories prevalent in this area. Of the experiments in which I participated, one in South Wales started with the exact weighing out of three dozen organic lettuce seed batches. The year after that (1978), a very intensive experiment was performed, in which four rows of radish were sown each day for well over a month. It was quite a few years later that I acquired the skills to be able to analyze this data at all adequately. I did not sow the rows in these experiments, but had a part in their design, and sometimes helped with the harvesting.

Amongst gardeners who plant by the moon, there presently exist two differing and incompatible traditions, of perhaps equal popularity. This is the reason why an experimental approach is required. One flourishes chiefly in America, but is also widespread throughout Europe and no doubt the rest of the world, and is based on the Tropical zodiac, tending to recommend use of only one-quarter of the zodiac, namely its water signs, for the sowing of crops. The other tradition which has its source in middle-Europe is based on Maria Thun's work, using the Moon's journey through the constellations as the basis for work on the land. It uses a visible scheme of things.

There is a second reason why a work focussing on the experimental evidence may be called for at this period of time. Within the Biodynamic movement itself, the experts are notably failing to endorse the 'Thun' approach. From Dornach and Darmstadt, waves of deep skepticism have been emanating. Thun's calendar has been going now for over three decades, is presently translated into no less than twenty-one different languages, and each year sells in tens of thousands throughout Europe - though far less in Britain than other countries. Its sales have grown, as far as this writer has been able to ascertain, without advertising, i.e. its use has spread by one grower telling another about their experience using it. Yet, various works by (male) Biodynamic experts have been averring that it uses the wrong monthly cycles and that experiments by others have failed to confirm its core principles. Articles by the present writer have endeavored to set the record straight.[7]

The structure of a lunar calendar involves debate over the zodiac and constellation framework, a difficult theme postponed until Chapter Eight. It is a matter which does generate controversy. The present work aims to ascertain the principles for evaluating the link between plant and cosmos. If we do not enter into these subtle distinctions, then the confusion will continue. The reader may here gather, in a mere few hours of reading, matters that it took the writer twenty years to fathom. This chapter touches on contemporary debates astir within America and Europe.

The call is here made for an experimental and open-minded approach to the subject. Time will no doubt tell which of the traditions accord most with reality. The evidence gathered so far appears to suggest that neither of the existing traditions are using the most appropriate celestial reference, but that a system of gardening/farming based upon the unchanging splendor of the Sidereal zodiac, as defined twenty-five centuries ago by the Chaldeans, is more in tune with the cosmos. Use of the Sidereal or star-zodiac is in accord with the Space Age into which we are now entering. A theory which claimed to be based upon it could potentially be of quite widespread interest.

I keep waiting for an edition of the Archers where one character says, "We can't sow the potatoes today, the Moon isn't right". I guess the nearest thing to this happened in September 2009 when the Maria Thun book about wine-tasting emerged. It was suddenly revealed that major supermarket-chains were sampling their wines using the correct Moon-trigon! Various sober experts were cited in The Guardian and BBC as totally validating the experience. It wasn't science, some complained; maybe not, but it was experienced. Suddenly the ancient concept of the Moon journeying in front

---

7 N.K. and Gerhard Staudenmaier: 'Mond-Trigon-Wirkungen Eine statistiche Auswertung' Lebendige Erde 1998 November 478-483, English translation in 'Harvest, New Zealand Journal of the Seasons', May 1999 (These are both Biodynamic journals), and 'Evidence for Lunar-Sidereal Rhythms in Crop yield: A Review' Biological Agriculture and Horticulture, 2001, 19, 247-260: (PDF).

of the Ram, the Lion and the Archer acquired a new relevance - amongst wine-connoisseurs. If there is an imbalance in the book you are about to read, it is that it too much emphasizes quantity, what can be measured, rather than quality, viz. the taste. And wine-tasting is the apotheosis of quality-testing! Let us therefore look forward to more such tasting parties, 'fruit versus root.' It may help us to appreciate more the cosmic nature of the fiery essence of wine. And, once again, we take our hat off to that great-grandmother Maria Thun for her ability to re-envisage things and to conceive of a truly simple experiment.

The original articles have been edited as necessary. I thank Simon Best, who has taken a similar down-to-earth approach to celestial influences, for assistance and for many of the references involved; also Robert Powell, for his explanations concerning the constellations and zodiac. To those who gave of their toil, soil and time, without thought of reward, I dedicate this book: Reg Muntz and Colin Bishop. May their work be a beginning.

A debt of gratitude is due to the women who dedicated their lives to working with cosmic influences in agriculture: Lilly Kolisko, Agnes Fyfe and Maria Thun. Their work combined intuition with science, reached between theory and practice, between sky and earth, as is necessary for understanding the living realm of Time in Nature, to help establish the ecological science of tomorrow.

I am also grateful to RM, for encouragement. Tactfully he explained that, no, I couldn't just reprint the collected articles, but that they had to be re-cast and synthesized. I came to apprehend how to analyze the waveforms involved. While attempting to resolve the different components present in the data, diurnal and monthly, solar and lunar, terrestrial (geomagnetism) and celestial, the notion of the Silver Axioms began to unfold.

1. Perspectives

The precious things put forth by the Moon - Deuteronomy 33:14

A lunar-gardening or farming calendar contains predictions for the forthcoming year. It aims to discern the future, whereby some days will be fertile and others infertile, etc. It is the business of science to make predictions about the future, but the predictions here involved tend to make scientists uneasy.

This is because they are qualitative, averring that the quality of time differs from one day to another, and does so in a periodic and cyclical manner. The predictions are testable, as claims are made for one period as being beneficial while another is adverse. If there were not such testable differences then a calendar would be mere hot air.

Today there is a need for sciences more concerned with the web of life than with analysis of its parts, with the overall ecology and balance of the whole, of which we form a part. This must involve an understanding of how cycles of time work within living organisms.

For the farmer, this could involve more of a role for the faculty of intuition, in assessing the cosmic and earthly factors that are relevant to a decision. A farmer deciding when to sow or harvest his crop has to ruminate on the matter. No computer can tell him the answer, and a wrong decision can cost a season's crop. The onions will not dry properly, or frost will reach the newly-sown potatoes.

In the past, scientists have been rigorously trained to eliminate their feelings and intuition from the setup as far as possible, and from the writeup of the experiment. People are now becoming uneasy about this dehumanised approach, and have come to blame it for much of what is wrong in the modern world.

We need today a more subjective approach to the study of Nature, and a more constructive use of factors hitherto dismissed as 'astrological'. Let us here take a historical example.

Aristotle was addressing the question, of what use is philosophy? A modern-day academic would be hard-pressed to find any reply at all to this question, however Aristotle had an answer. It went as follows:

> It would be well also to collect the scattered stories of the ways in which individuals have succeeded in amassing a fortune; for all this is useful to persons who value the art of getting wealth. There is the anecdote of Thales the Milesian and his financial device, which involves a principle of universal application, but is attributed to him on account of his reputation for wisdom. He was

reproached for his poverty, which was supposed to show that philosophy was of no use.

According to the story, he knew by his skill in the stars while it was yet winter that there would be a great harvest of olives in the coming year; so, having a little money, he gave deposits for the use of all the olive-presses in Chios and Miletus, which he hired at a low price because no-one bid against him.

When the harvest-time came, and many were wanted all at once and of a sudden, he let them out at any rate which he pleased, and made a quantity of money. Thus he showed the world that philosophers can easily be rich if they like, but that their ambition is of another sort.[1]

Thales of Miletus predicted a bumper harvest of a specific crop for the forthcoming year, according to Aristotle. Let us hope that we may in some measure discern the 'principle of universal application' which he used to accomplish that prediction. This helped to confer upon Thales his reputation for being not just well-informed, knowledgeable or clever, but wise.

The Roman Pliny the Elder wrote an extensive 'Natural History' in which the lunar traditions employed by the practical, hard-headed farmers of the Roman Empire were described. He there explained that the 'deep question' of the "fit time and season of growing corn ... would be handled and considered upon with exceeding great care and regard; as depending for the most part upon astronomy." He alluded to the above story of Aristotle's about Thales as an example of what was possible.[2] Pliny was convinced that a right understanding of such matters, if only it could be achieved, would be of the utmost practical value:

> It is an arduous and vast aspiration, to succeed in introducing the divine science of the heavens to the ignorance of the rustic, but it must be attempted, owing to the benefit it confers on life.[3,4]

His 'arduous and vast aspiration' remains an unfulfilled promise - and today we may wonder, from the mulch and well-rotted compost of old

---

1 W.D.Ross, ed., 1928, The Oxford Translation of Aristotle, O.U.P. Dr Lee Lehman pointed out this quotation and its relevance.
2 Pliny, *Natural History*, Book 18, p.363 (Thales story): 'Such is the opportunity afforded by learning, which it is my intention to introduce, in treating of the operations of agriculture, as clearly and convincingly as I am able.'
3 Pliny, *Natural History,* Volume 18, sections XVII-LXXV,LVI, e.g.: 'All cutting, gathering and trimming is done with less injury to the trees and plants when the Moon is waning than when it is waxing. Manure must not be touched except when the moon is waning, but manuring should chiefly be done at new moon or at half moon. Geld hogs, steers, rams and kids when the moon is waning. Put eggs under the hen at the new moon...In damp land sow seed at the new moon and in the four days round that time.' p.391.
4 'The Roman Farmer and the Moon', *Transcripts and Proceedings of the American Philological Association,* 1918, 49, 67-82.

superstition and folklore, is the fair bloom of a new science ready to emerge?

Kepler's View

Johannes Kepler, one of the founders of modern science, first rose to eminence as a composer of almanacs. In one year (1595) his calendar predicted a severely cold winter and an invasion by the Turks, both of which came to pass. History books record his more enduring fame in discovering the planetary laws of motion, but the locals were probably more impressed by his yearly prognostications.

Kepler affirmed the practical utility of such matters, opening the Preface to his 1602 Calendar with the words:

> Discover the force of the Heavens O Man; once recognized it can be put to use. That of which we are ignorant can profit us nothing. Only futile labour is onerous; success brings gain. By your skill, O Men, discover the force of Nature.[5]

The calendars spanning three decades of Kepler's life (1595 - 1624, of which only half a dozen or so remain) were mainly written in old German, and none of them have as yet been translated, a situation which regrettably shows no immediate signs of changing. He composed one calendar in Latin for the year 1602 which has been translated (above-quoted).[6]

Its predictions were based upon planetary aspects, with the Moon entering into consideration chiefly via its phases, and the planetary aspects gathered around its Full position each month.

Kepler's preface to his 1602 calendar described how a composer of calendars ought to consider the limits of what can be predicted in advance, avoiding on the one hand undue fatalism and astral determinism, and on the other hand excessive timidity and caution - a perilous Scilla and Charybdis one was required to steer between.

A contemporary calendar composer had gone into undue detail about how the wine harvest was going to turn out in the coming year, using a procedure of which Kepler did not approve, and he described such opinions as "the stuff that empty dreams are made of." Hoping that we shall avoid meriting such a dire accusation, we here endeavor to develop a Keplarian approach

---

5 Kepler, Frontispiece to 'A physical Prognosis for the coming year 1602' Kepler's Preface to that calendar was entitled *De Fundamentis Astrologiae Certioribus*, 'On giving astrology sounder foundations' (Field, 1984, p.229).

6. I discuss this in 'Kepler's Belief in Astrology', in *History and Astrology, Clio and Urania Confer*, Ed. Kitson, 1990:. The translation of Kepler's 1610 *Tertius Interveniens* 'third man in the middle' i.e. in-between astrology and astronomy – translated as 'Kepler's Astrology' by Ken Negus in 2008, has much about what kinds of prediction he viewed as being legitimate for his yearly calendars, and how it all worked. 'Kepler's Astrology' by NK and Nick Campion, *Culture and Cosmos* 2012, has a translation of his 1618 ephemeris.

to the subject. Kepler presented an early version of the Gaia hypothesis, explaining how; "the Earth has a vegetive animal force, having some sense of geometry."

In a beautiful analogy, he explained how, just as a peasant could take delight in the piping of a flute, without having any knowledge of musical harmony, so likewise the Earth responded to the changing geometry of the heavens:

> ... that power which makes the aspects effective must be inherent to all sublunar bodies, indeed the whole Earth. The entire vital power is, you see, a reflection of God, who creates according to geometric principles, and is activated by this very geometry or harmony of the celestial aspects.

Thus, he explained how;

> Eclipses are so important as omens because the sudden animal faculty of the Earth is violently disturbed by the sudden intermission of light, experiencing something like emotion and persisting in it for some time.

Our prime concern here will be with the question of evidence: can it be shown that eclipses have any such effect upon living things? The belief that eclipses have a rather substantial and negative effect upon the land reaches back for a good two and a half millennia, being probably the oldest continuous tradition.[7]

Later works of Kepler discuss more by way of human astrology, and of course astronomy, but his early work of 1602, from which we have quoted above, De Fundaments Astrologiae Certioribus ('On the more certain fundamentals of astrology', or, 'On giving astrology sounder foundations') is of especial relevance for us as pertaining to the manner in which the Earth responds to celestial influence. It was composed in his 29th year, during his Saturn-return (i.e. as Saturn returned to its position at his birth), as he was about to inherit the post of Imperial Mathematician from Tycho Brahe. It is primarily about matters relevant to the construction of calendars, and not with personal horoscopes.

There presumably exists a continuous tradition from the calendars of Kepler to the modern Biodynamic calendars of middle-Europe, however it is well buried and likely to remain so for a while yet. There are many histories of astronomy, and a few of astrology, but none of the prolific gardening-almanacks and annual calendars of the kind that interest us. Though enduring through millennia, the tradition has hardly concerned historians. Even attempting to discern the eight decades of history of the

---

[7] For the ancient Babylonian belief that a solar eclipse 'damages the seeds of the earth,' see Michael Baigent, 'From the Omens of Babylon, Astrology and Ancient Mesopotamia' 1994, p.103.

modern Biodynamic calendar has been far from easy, depending upon locating a few elderly persons who could in some degree recall matters. The account given here is incomplete, but - hopefully - better than nothing.

Nowadays we inhabit a vastly different intellectual universe from that which nurtured such beliefs, where space reaches into black holes and military hardware tracks across the skies. The physicists speak a strange esoteric language if one is rash enough to ask them about influences of the heavens, more appropriate for discussing ozone holes than the nurturing of cabbages. Still, 'Why seek ye the living amongst the dead?' We are concerned with the life-energies to be found in the vegetable realm, for which little help may be available from sciences oriented towards a non-living, mechanistic realm.

Can we hope to really 'discover the force of the heavens' such that it can be 'put to use'? For three centuries science has been assuring us that such notions are mere old wives' tales, and that we should shape up to living in the real world, where science takes a more objective approach. This has led us into a world of objects, where things can be taken apart, flowers genetically manipulated and star positions measured to a millionth of a degree. But, I may object to paying taxes to support such research and prefer to listen to old wives' tales. For three centuries men have been breaking up matter into smaller and smaller pieces, and the world teeters on the brink of destruction in consequence.

The word 'superstition' means, 'that which stands above', from the latin 'super-stition', whereby it represents an antithesis to the term, understanding, which is what one sees or appreciates from 'standing under'. One expects the latter to be earthy and solid, with 'concrete' proof etc, while the former will be more airy - and *linked to the stars above*. There is a sexual polarity here, with men (almost entirely) controlling the reality-concepts of a materialistic science concerned to 'understand' natural phenomena, while the important beliefs that concern us are smilingly dismissed as 'old wives tales'.

We could here distinguish between what is understood and that which is considered - the latter meaning in its derivation, 'together with a star' (from the Greek, sidera, a star, and con- together with), i.e. an act of contemplation. On this view, to map the gene structure of a plant would be to gain understanding of its function, while to discover patterns in the time-structure of its growth is to consider it.

Plants have larger DNA coils in their cell nuclei than do animals or humans. We review evidence of how the character of DNA in plant cells varies with the monthly lunar cycle (Ch.4), which may tell us something about fertility as inherently linked to that cycle. The large volume of plant DNA enables such fluctuations in volume to be detected quite readily.

So alienated has our culture become from the cosmos, that only the military in post-war Britain have utilized the starry vault. In my youth, the deadly missiles of the submarine 'Polaris' were designed to rise above the stratosphere where they would orientate themselves briefly with respect to the Pole Star before plunging on their downward path. No-one else was concerned - astrologers talk endlessly about the stars, 'What the stars foretell' etc, but don't ever use them. Their zodiac came unlinked from the stars about fifteen centuries ago, with the arrival of the Dark Ages. More recently, there was a British school of 'sidereal' astrologers, that became extinct several decades ago. Undeterred, we seek for what the poet called; "the starry dynamos of night."[8]

The pioneer scientists described in this opus are women: Lili Kolisko, Agnes Fyfe and Maria Thun. It is they who mapped out the relevant time-patterns, relationships between earth and heaven, of practical use for gardeners. The future needs a more feminist science, concerned to investigate the matrix of interrelationships on which life depends and less concerned with the breaking up of matter.

Fertility

> ... the female functions, although they do not coincide in time with the Moon's phases, coincide with them in their periodicity... It is as if this process of the female function were lifted out of the general course of Nature, but has remained a true image of Nature's process. - Rudolf Steiner[9]

Once a month, a woman becomes fertile. Women living together may find that their cycles tend to synchronize, although this was not recorded at all in the literature until 1971.[10] More recently, US surveys have established that women whose periods are of quasi-lunar length, i.e. between 28 and 31 days, tend to have their cycles phase-linked to the Moon.[11,12] The

---

8 Allen Ginsberg, *Howl and Other Poems*, San Francisco 1956.
9 Rudolf Steiner's 1921 'Astronomy Course' lecture 2, translation by Olive Whicher.
10 Martha McClintock, 'Menstrual Synchrony and Suppression' Nature 1971, 229, 244-5.
11 Winifred Cutler, 'Lunar and Menstrual Phase Locking', American Journal of Obstetrics and Gynecology, 1980, 137, p.834. For a replication, see, in the same journal, letter from Dr Erika Friedmann (1981, 140, p.350). But, a 1982 study conducted on 826 young female volunteers in Beijing, China found that twice as many menstruations occurred during the four days centred on new moon, as over the Full. This is a phase-reversal compared with the Cutler result, possibly reflecting a less industrialised condition of society. (Law 1986, cited by Chris Knight, 'Blood Relations, Menstruation and the Origins of Culture' YUP 1991, p.249).
12 For a thorough review of the evidence see Alexandre Dubrov's, Human Biorhythms and the Moon (1996), pp.57-68. While Dubrov's magnum opus, The Geomagnetic Field and Life

phenomenon continues to be ignored by doctors and psychologists, who have successfully deluded womankind into regarding the monthly period as averaging 28 days in length. A period of four weeks is convenient for schedules of contraceptive pills, but large-scale surveys indicate that its mean length is 29½ days.[13,14,15] Womankind has been deprived of its cosmic birthright for the convenience of pill-packet instructions. This enabled medical experts to evade the awkward question of what the lunar cycle is doing in the reproductive processes of womankind, while being more or less absent in other mammals.

This cycle has a responsiveness both socially, ie between women, and in relation to the meetings of the two luminaries in the sky. Biologists hop to the conclusion that the hormone changes linked to this cycle are themselves the clock which times it, as estrogen and other hormones fluctuate with the monthly period. However, that is not really self-evident.[16]

The period of gestation averages just nine lunar-month cycles, underlining the significance, the uniquely human significance, of this period for birth, fertility and growth:

$$9 \times 29.53 = 266 \text{ days.}$$

Nine times do the Sun and Moon meet in the sky as a child grows within the womb. Large-scale surveys enable us to cite those numbers with a fair degree of confidence. Surveys sometimes come up with a figure one or two days shorter than the figure of 266, as the overall mean estimate for the duration of pregnancy. Of interest here are those which have resulted from artificial insemination, as giving a definite date for conception, which is otherwise hard to come by.[17,18] My feeling is that the data is now precise

---

was translated from the Russian, this has evidently been composed in English. It introduces much Russian research to the English-speaking world.

13 A. Treloar et Al., 'Variations of the Human Menstrual Cycle through Reproductive Life', International Journal of Fertility, 1967, 12, p.77-126.

14 P.H.Jongbloet, (Holland), *Dev. Med. Child Neural.*, 1985, 25, p.527: '...the menstrual cycle in women does not average 28 days. Both its mean and median length are 29.5 days...' Its modal, ie, most frequent length is around 27 days, as the distribution is quite skewed: Dubrov, 1996, p.66.

15 The most exact study has been that by Dr and Mrs Vollman (R.F.Vollman, 'The Menstrual Cycle' 1977, Vol. 7 in *Major Problems in Obstetrics and Gynecology*, Ed. Friedman). Covering some thirty thousand menstrual cycles in over six hundred women of all ages, they found the mean period to be 29.5 days, while (as above) the mode i.e. most common period was of 27 or 28 days. Vollman had no interest in lunar lore.

16 F.Brown, 'Biological Clocks: Exogenous Cycles Synchronized by subtle geophysical rhythms' *Biosystems* (Amsterdam) 1976, p.68.

17 R.D.Martin, 'Female Cycles in relation to Paternity in Primate Societies' in Martin et al., *Paternity in Primates* Basel 1992, pp.238-74.

18 There is a suggestion from *in vitro* fertilization studies that the mean human gestation period is a few days shorter than the 266 day estimate. One can take the view that the mean gestation period more resembles 263 days, this being the figure usually cited for the duration

enough to confirm the original view advanced by the Menakers that the period is a function of the synodic month, and that its mean value is 9.00 such months.[19],[20] It does not support the older (astrological) tradition of ten sidereal months, which would give a gestation period of 273 days. The notion that there are ten months in gestation is surprisingly widespread but is, so to speak, a misconception.

Women notice that they tend to synchronize their cycles with their closest friends, even if they are not living together. In this there can be a dominant partner and, if the cycling of her period suddenly shifts, as does happen from time to time, the other will follow. It is becoming OK for women to acknowledge these things nowadays, even if not for biologists to discuss them. The latter tend to dismiss the subject with a nervous laugh, as if frightened that their theories would fall apart if they started to take such things seriously; as they probably would.

The lunar-fertility link in womankind goes beyond the mere coincidence of cycle length referred to above. Firstly, as we have seen, the 'lunar-period cyclers,' roughly one-third of the population of menstruating females, whose periods range between 28-31 days, will tend show some degree of synchrony with the meeting of the luminaries in the heavens[21]; secondly, populations of women indulging in regular sexual activity have a larger proportion of cycles approximating to 29.5 days (the lunar month), than in groups of women for whom this is not the case[22]; and thirdly, the childbearing years are those when the period length most closely approximates to that of the Moon: earlier, in the teens, it is slower and can average thirty-five days or so, then in the thirties its average is around twenty-eight days and in the forties it decreases to twenty-seven days. These are overall means. So the length of the female period tends gradually to decrease with years. These are matters deserving further study.

In gathering the horse data from Newmarket stud farms, I discerned some degree of synchrony in the estrus cycling, which for horses is three weeks. Three-week waveforms were running through the covering data ('covering' refers to the mating process), unknown to the vets. This discovery was fortunate, in legitimizing my presence there. It was acceptable for such

---

of Venus's visibility as either Morning of Evening Star. Thus the two traditionally feminine planets, Moon and Venus, inscribe in the heavens this uniquely-human period.
19 Walter and Abraham Menaker, 'Lunar periodicity in human reproduction: a Likely Unit of Biological Time' American Journal of Obstetrics and Gynecology, 1959, 77, p.905-914.
20 . T. Criss and J. Marcum, 'A Lunar effect on fertility' Social Biology 1981 28 pp.75-80.
21 See Cutler, ref. 11.
22 Chris Knight, ref 11, p.248: "only about 28% of reproductively active women show a 29.5 ±1 day cycle length ... cycles of this length tend to be the most fertile ones. The finding that there is a positive correlation between fertility and precision of lunar phase length has been described as 'an intriguing biological coincidence' (Cutler et al. 1987)". Knight here alludes to the Treloar and Vollman surveys (refs 13,15).

synchrony to exist, in a way that a lunar-fertility study was not acceptable. It was now and then present between stud farms a couple of miles apart, which was more problematic: were pheromones drifting downwind, perhaps? I suggest that such synchrony, ie coming on heat together, is natural, ie normal in the course of nature, even if only perceptible in horse data where all coverings are carefully logged.

In horse fertility, three- and four- week rhythms are interacting all through the spring and summer months. The former rhythm is endogenous, the latter exogenous. Endogenous means 'generated within', like a heart-beat, ie not given by the cosmic process, so that no external clock is timing it. The three-week rhythm is endogenous because it belongs to the horse, not to the Earth or the heavens. The horses tune in together by synchronizing these cycles, so that they come on heat together. Why do they do this? I tend to query whether such a question needs answering. Rhythms exist in life-processes at every level. It is the nature of living things to have synchronized cycles. In the future we should be able to write the fertility-equations, for predicting the periods of optimal fertility and even, it may be, avoiding periods when birth failures are more likely.

Primates range in their estrus cycle periods from seven to thirty-nine days. A recent study gave a list of twenty-five 'cycle lengths of non-human primates'. The chimps had 37 day cycles, some gibbons had 30-day cycles, and baboons had 31-35 day periods.[23] The overall mean of all these periods, cited to within a day, I found to be 29 days. One is here reminded of monthly cycle lengths for womankind, whose periods vary widely, so that the figure of 29 or 29.5 days only emerges as the overall mean.

Was there a period in human evolution when the lunar cycle of womankind was collectively synchronized to the lunar cycle? Here we enter realms of giddy speculation, but not without relevance to the concept of fertility. An anthropologist has argued the case for such.[24] Oestrus and thus fertility would then have been collectively experienced as a Full Moon event, and menstruation as a New Moon event. There seems to be little direct evidence in support of this theory, but a vast amount by way of mythology, tradition and comparative anthropology that makes sense from assuming it. The linkage of these two ends of the monthly cycle with the female experience has a significant analogy in the plant realm, if seeds indeed tend to germinate and grow optimally around or prior to the Full Moon.

Claudius Ptolemy, who was the astronomical, astrological and geographical expert at Alexandria in the century after Pliny, synthesized

---

23 Knight, ref (11), p.248. See also F Brown, 'Common 30-day multiple in gestation time of terrestrial placentals,' Chronobiology International, 1988, 5, pp.195-201, showing that the lunar month or sub-mutiples of it were common in estrus-cycle periods.
24 Knight, ref (11), Ch 10, 'Hunter's Moon.'

much of ancient knowledge in his *Tetrabiblos*. Reviewing farming practices of his day, he clearly affirmed that:

> Farmers notice the aspects of the Moon, when full, in order to direct the copulation of their herds and flocks...and there is not an individual who considers these general precautions as impossible or unprofitable.[25]

Today this is a largely-forgotten topic, such that it would be hard to find any farmer who had heard of such beliefs, let alone practiced them. And yet, the ancients took it for granted that the male-female duality was encoded into the cosmos, as in the difference between the Sun and Moon, or in the contrast between their qualities of 'dry' and 'moist'. It seemed natural to them that fertility should vary with the course of the month.

When Korringa discerned, in the 1930s, the lunar key to oyster swarming periods, as used since by Dutch fishermen, he grew rather apprehensive.[26],[27],[28] A crisis had arisen amongst Dutch oyster farmers, due to the rather sudden peak in the swarming of oysters, and they would be gone before the nets were ready. Korringa observed that the queen oyster was attuned to the Full Moon of the breeding season. It was, he pointed out, a certain number of days after a specific Full Moon in June or July, that huge swarms of these creatures could be caught by the fishermen, following their spawning during the Full Moon. His article kept remarking that he was fortunate he did not have to worry about the old laws against witchcraft. Today the climate of opinion is shifting, with the development of ecological attitudes.

Reacting against today's mechanized and robotic society, people are seeking for more organic modes of thought.

Sidereal Moon

As well as the monthly cycle of the synodic period and its deep connection with fertility, we are here concerned with the starry or 'sidereal' month of twenty-seven days. In this time, Luna orbits the Earth. The differing functions of the synodic and sidereal months intertwine as core themes of the present work. There was a time, before the idea of the zodiac was formed, when people recorded the Moon's passage against the

---

25 Claudius Ptolemy, Tetrabiblos, Ch.1.3.
26 P. Korringa, 'The Moon and Periodicity in Breeding of Marine Animals,' Ecological Monographs, 1947, 17, pp.348-381.
27 For a similar effect, see 'Lunar phasing of the Thyroxine Surge Preparatory to Seaward Migration of Salmonid Fish' Science, Feb. 1981, p.607-9, on how salmon have their migrations triggered by a hormone surge at New Moon.
28 For a detailed review of earlier studies of organism adaptation to the lunar cycle, see Caspars, 1951; its references alone occupy twenty pages; and more recently, Endres and Schad, 2001.

constellations of the night sky.[29] Our theme takes us back to that primitive, primal, experience. What does it mean to use just twelve constellations on the ecliptic (the line through the middle of the zodiac)? Modern astronomers map thirteen constellations as lying across the ecliptic - plus some more that lie within the band of the zodiac. Perhaps we should compromise with twelve and one- third ecliptic constellations, as one of these constellations is only partially present athwart the ecliptic, by analogy with the twelve and one- third lunar months per solar year. Chapter Eight reviews man's evolving experience of the zodiac as it is relevant to work on the land.

The 'Planting by the Moon' calendar published through the 1980s had the distinctive feature, in that it used the original sidereal zodiac, as was used in antiquity and formed in Chaldea. It used twelve equal signs. This can generate far-reaching confusion over the terms 'sign' and 'constellation'. The latter pertains to stellar images, as may or may not lie on the ecliptic, of unequal size. Are there twelve or thirteen constellations on the ecliptic? Use of the term sign however must pertain to an equal twelvefold division, with each covering thirty degrees of the ecliptic. In what follows, differing views will appear as to whether a lunar gardening calendar should be based upon signs or constellations.

In the sidereal zodiac, the twelve equal signs represent a 'best fit' grid placed upon the unequal constellations, and stars define its position; whereas the zodiac used by today's astrologers is strictly linked to the Sun's position at the Vernal Point. It has no connection whatsoever with the stars, and as such it is moving ever further away from the constellations which gave it birth.

In the beginning of 1995, in the British newspapers, the stars made headlines, as the UK's Royal Astronomical Society proclaimed that there were thirteen constellations on the ecliptic. This was proclaimed as some sort of discovery and a fine opportunity to knock the astrologers, as happens from time to time. The 'thirteenth constellation' turned out to be Ophiucus, the serpent-bearer, traditionally overlapping with the Scorpion. Its foot enters right into the zodiac and steps onto the Scorpion! A historical perspective is necessary to have any hope of avoiding confusion over this matter.

We here reclaim an experiential, sensible approach, concerned with what is perceptible to the senses. We remain with the stars visible in the sky, as our local region of the galaxy. From our geocentric perspective they form patterns that are humanly important. Around the zodiac, where the planets travel, they delineate the energy patterns. Modern astronomy has a galactic or extra-galactic perspective, going far beyond the stars visible to the eye. Astronomers gaze into computer screens at strange objects beyond the

---

29 Christopher Walker Ed, Astronomy Before the Telescope, 1996, p.49.

galaxy, then complain that the public aren't interested. Instead, by way of developing a more experiential view of this subject, books by Davidson and Schultz[30] may be recommended. A modern astrological authority, Nicholas Campion, has expressed the view that:

> The Biodynamic system is thus a powerful argument in favour of the use of organic farming methods and a pointer to the most positive help which astrology can offer.
>
> Out of all the uses of astrology, the regulation of gardening and agriculture according to cosmic cycles has the greatest practical potential.[31]

Can such a potential indeed be realized - and if so, would it be 'astrology'? Astrologers would be reassured if their art could be seen as connected with the rest of nature. Quoting from Mr Campion's conclusion:

> "It is often said that astrology is a luxury for those with time to dwell on their own problems. Yet if astrology can help to feed the starving there can be no greater reason for the spread of its study and use."

Biodynamic farmers would tend to be uneasy about that compliment from an astrologer, regarding their own calendar as a separate and independent enterprise. Yet, such a point of view is rather isolationist: there is, after all, only one sky above us, and whatever influences hold sway through the patterns of time should be of equal interest to both camps.

Through use of a lunar gardening calendar one can develop a relationship with the heavens of our solar system. The farmer acquires a framework for scheduling the week's work, and a motive for looking up to the stars at night. Agriculture acquires a new dignity. The farmer is empowered - and the fishermen catch their oysters.

Organics

The younger generation will presumably never know the drowsy, murmuring sound of bees in summertime, as instead they see only the odd lonely bee buzzing around. This happened in the 1990s: somewhat as we might wish to consult our grandparents about the sound made by a flock of swans flying by overhead, which they used to see. What will fertilize the peas and the apples once the bees are gone? One shudders to think of the solutions which modern agribusiness is likely to come up with. No-one can have organic bees in the UK, as pesticide spraying is sufficiently

---

30 Norman Davidson 'Astronomy and the Imagination' 1985; Joachim Schultz, 'Movement and Rhythms of the Stars' 1963 (English translation by John Meeks, 1986). The latter is more difficult, and probably less relevant to the present work.
31 N.Campion, Ch.22, 'Planting by the Stars - Astrological Gardening,' The Astrology Book, 1990.

widespread to infect most bee colonies. What is most annoying here is that the persons responsible are never going to shoulder the blame: those who have so undermined bee vitality by the poisons sprayed onto flowers just shrug their shoulders and point to an object, say the varroa mite, as the culprit. It has come over from France. Is not the attitude, of just wanting to optimize production and taking too much honey from the hive, over decades, what has led to this catastrophe?

A veritable bee-apocalypse faces farmers in Europe and America as colonies of bees are just vanishing. Some surmise that mobile phone masts are disorienting the bees, others say that bee-autopsies point to new and more virulent pesticides, some associated with GM foods, as the problem.[32] Thus an urgent wake-up call comes from Mother Nature, concerning the absolute need for a biological and organic way of agriculture. Experts call for reservoirs of organically-farmed land to enable bees to survive.

A statement by the UK's Institute of Science in Society advised how sub-lethal levels of various insecticides were sufficient to disorient the bees and prevent them from ever returning to their hive. If the bees aren't there to pollinate flowers:

> Most fruit and many vegetables would disappear from our diet along with an immediate shortage of meat, due to the loss of forage. ... There are scientific studies showing that agricultural landscapes with organic crops are far superior environments for both honey- and bumblebees. [33]

British farmers today have the highest suicide rate of any profession. Even in the early 1990s, prior to the Mad Cow epidemic, their suicide rate was 50% above the national average.[34] Dismally, it is now the second commonest form of death amongst farmers up to 44 years old. By way of a remedy, we here advise farmers to put aside for a while their ingrained axioms about progress and efficiency, and re-evaluate matters. The poisons which they spray so liberally upon Mother Earth, which kill the little creatures, make them disliked by the public. In the words of the Bible:

> *There is a way that seemeth right to man, but the end thereof is death.*   - Proverbs 14:12

The straight lines through the cornfields, left by the tractors as they spray the pesticides, symbolize the linear logic that has led to such dreadful error. Farmers have only done what they have been told, and now more than one

---

32 Joe Cummins, Saving the Honeybee Through Organic Farming (Presented at launch conference for 'Food Futures Now *Organic *Sustainable *Fossil Fuel Free' , 22 April 2007, UK Parliament, Westminster, London)
33 Ibid.
34 Liz Mason, 'Stress takes a Heavier Farm toll', Farmer's Weekly, 20 November,1992 10-14.

in three are unhappy or depressed, and one in five have experienced stress-related illness per year, a far cry from the traditional rustic idyll.

Let's cite a humdrum example, of how to deal with carrot-fly, a problem faced by every vegetable-grower. Maria Thun's solution involves time:

> Carrots like mature soils that have not been manured for two years ... While late manuring detracts from the taste of carrots as far as we are concerned, it makes them more attractive for the carrot fly. It finds carrots cultivated like this ideal as nurseries for raising its young ones ... Carrot flies are still keener on soils to which peat has been added, since they then find it easier to reach the roots and lay their eggs.[35]

Usually growers put onions and garlic next to their carrots, while covering with some gauze to further repel the insect. This different approach advocated by Thun may be worth a try. Her book integrates down-to-earth advice with celestial guidance in a way that has deep roots in Germanic folk-tradition.

At the yearly organic food trade fair at Nuremberg, the Demeter and Soil Association labels feature prominently. Organic methods of farming are increasingly able to compete with chemical-intensive methods in the marketplace, as the public become more apprehensive about food-scares and the poisons used by conventional methods. In the case of cotton, for example:

> For the first few years after being introduced, organic farming was less productive than chemical-based production. But with the right type of biotechnologies and perseverance, the yields can match those from pesticide- and herbicide-dependent systems. In 1994 the world market offered $2.77 per kilogram for organic cotton as compared to $1.32 for conventional producers - an unheard-of premium in the world of markets of agricultural produce. And as demand from the fashion designers continues to grow, there is a guaranteed market in which to sell.[36]

Organic coffee used to have its premium price in excess of 100 per cent, but this is now below 30 per cent.

Nearly half of organic farmers in Europe are women, either as equal partners or farmers in their own right, which seems rather significant. In Britain, despite fifty years of endeavor by the Soil Association, only about one percent of farmers use organic methods, and in the UK and in Europe as a whole organic farms occupy 4% of the farming area.[37] Denmark,

---

35 M. Thun, *Gardening for Life*, 1999 p.86.
36 Gunther Pauli, *Breakthroughs, What business can offer Society*, 1996, p.129.
37 http://www.organic-europe.net/

Finland, Sweden and many German provinces presently grow 5- 10% of their food organically. Britain's struggling organic growers are indignant that huge subsidies continue to go towards chemical-intensive methods. Also, they are livid at having to compete in a 'free-trade' area with Euro-organic farmers who get subsidies from their governments. Despite these drawbacks, the practice of organic farming is surging in the UK, as indeed it is worldwide.[38]

An article published in Science compared pairs or sets of Biodynamic and conventional farms. These were representative farming enterprises in New Zealand. In six of the seven farm sets it was found that the Biodynamically-farmed soils had better structure and broke down more readily to a good seedbed than did the conventionally-farmed soils. Organic matter content, soil respiration and mineralizable nitrogen were significantly higher on all the Biodynamically-farmed soils than on the conventionally-farmed soils. Earthworms were counted on the two market gardens to give another indication of biological activity: by mass, the Biodynamically farmed soil had 86 g of earthworms per square meter, whereas the conventionally farmed soil had 3 g of earthworms per square meter. It was found that greater organic matter content and biological activity contributed to the formation of topsoil at a faster rate on the Biodynamic farms.[39]

It is not the aim of this treatise to promote Biodynamic farming. I'd be far from competent to do any such thing; however, it is this system of farming and no other which is known worldwide for its use of a lunar calendar, and it is therefore vital for us to see that such farms do really work – those who run such farms are not just dreamy idealists with vaporous notions of 'muck and magic,' as one finds alleged in certain quarters.

In France and Germany, about two and five percent respectively of farmland is presently run on organic principles, some one-tenth of which is Biodynamic. About 1% of French wine is presently from grapes grown Biodynamically. These figures give some indication of how widespread the B.D. approach is, a subject on which figures are quite hard to come by.

The central, key fact to remember in any debate is, that organics gets more from the land than does intensive farming. It gets a higher yield per acre. Intensive farming is more profitable, for the few who make the profit, only because machines are doing the work - however, the per-acre yield is lower.

---

38 As a data-source, see here. The Sunday Times ('Mystic farmers get grants for zodiac growing,' 28.4.02) reported that a 'mystical form of agriculture' (i.e., Biodynamics) was producing wines that 'regularly win top prizes,' and that the number of these farms in the UK had doubled over the last four years.
39 J.P. Reganold, et al., 'Soil quality and financial performance of Biodynamic and conventional farms in New Zealand', Science 260:344-349. 1993.

For a farm to go organic may involve a drop in productivity, because that method is more labour-intensive. Young people hanging around the streets should instead be tossing bales of straw, collecting eggs and making hedgerows. Such activities are fun, community-building and promotive of mental and bodily health.

Algol & the British Farm Animal Holocaust

Not one cow born and bred as organic suffered from BSE (mad cow disease) in the 2001 epidemic, and hardly any were found to suffer from the foot-and-mouth outbreak. No foot-and-mouth was diagnosed on any Biodynamic farms in 2001, just as no BD farms were found to have it back in the more localized epidemic of the 1960s. As this work has a somewhat astral theme let us note that the world's worst ever foot-and-mouth epidemic hit the headlines in the last week of February 2001, just as Saturn was coming into conjunction with the evil star Algol. Algol the 'Medusa's Head' has the reputation of being the most evil star in the firmament and Saturn is traditionally the planet of agriculture. Saturn remained within one degree of Algol through most of March as the dreadful cow-burning began. A ghastly pall of horror settled over the British countryside as the massive extermination project began. Farm after farm was slaughtered-out. Only a tiny, tiny proportion of the cows slaughtered were ever diagnosed as actually having had the disease. A good six million farm creatures were slaughtered in response to a mild, nonfatal illness. The government's behavior exactly resembled that of a bodiless head, which had become fully severed from the body of British farming. Its madly-deranged logic resembled the writhing serpents around the head of Medusa, Medusa whose head turned to stone those who gazed upon her. 'Townie' experts within London's Imperial College and Royal Society used computer models to guide them in terrorizing the countryside, overruling the advice of the real, recognized experts on foot-and-mouth who were advocating a very different course.[40] A Dutch outbreak of foot-and-mouth during the same period was coped with swiftly and sensibly and with few deaths, by inoculation. MAFF (the Ministry of Agriculture, Fisheries and Food) itself dissolved during the slaughter, as the burden of guilt and the horror of its own actions became too much to bear.

Both Saturn and Jupiter were in the Taurus constellation. They were separating, having been conjunct the previous year. A farmer is concerned with the land and finds fulfillment through working with the cycles of Time, with that which passes down through generations, and it is this which makes

---

40 Christopher Booker, Not the Foot and Mouth Report, a special investigation from Private Eye, Nov 2001.

his or her work inherently Saturnine. As Dr Steiner observed, the farmer's job is a meditative one. Any farmer, leaning against the traditional garden gate, can readily describe what a sensible policy towards foot-and-mouth would be: one allows the creatures to recover from the disease, as they do in a few weeks, then one uses the ten percent or so that did not succumb to it for future breeding - and that is quite apart from the question as to whether or not one reckons there may be a medicinal cure for it. I suggest that this collective trauma now receding into history be used to mull over the concept of astral influence.[41] By contrast, there is a (traditionally) beneficial, first-magnitude star in Taurus: Aldebaran, the 'Bull's Eye' which is a pale-pink color, and perhaps medicinal or healing work on cows could consider use of this star.

The whole course of the disaster (N.B. *dis-aster*, from the Latin 'aster' a star) was supposedly caused by a small object, a 'virus.' Nothing about the cows mattered, except whether or not they had this object in their blood. It had miraculous properties such as traveling miles with the wind to re-infect new cattle. I am far from clear whether it ever existed, but that doesn't really concern us. If you ask a Biodynamic farmer about the impressive track record of BD farms (in the UK) of getting no BSE or Foot-and-Mouth, you'll quite likely hear a comment about them keeping the horns on the cows. Thus we have a different way of thinking about the problem.

The horns are the one part of a cow that reaches upwards. To remove them is to deprive a cow of its dignity. A farmer who leaves the horns on a cow, deserves respect from the community. There is something wonderfully mild about a cow, as if it had no idea of the harm it could cause with its horns. These days many city-folk don't even know that cows have horns! Removing them is quite a painful operation for the cow. From a more alchemical view, we could reflect upon the fact that the highest proportion of gold in the cow's body is found in the horns. This may remind us of the Egyptian image of the solar disc held between the horns of a cow, an image of glory, or again one may be reminded of the poet Virgil's line:

> *Glittering Taurus opened the year with his golden horns.*
> - Georgics, Book 1

- alluding to the constellation of Taurus as visible in the winter-time evening sky. Thus we apprehend that the balance of Chi or life-energy in the cow requires that the horns be left on.[42] Thus we are not merely

---

41 Algol, an eclipsing binary star, was at 26° 17' Taurus in 2001. For dire stories about Algol see the section in Diana Rosenberg's website Fixed Stars and Constellations or my web-article on Algol .

42 A 2003 German dissertation by Jennifer Wohlers studied milk quality within a single herd of partially de-horned Fresian cows, and concluded that the milk from horned cows had a superior quality. The latter was shown by tests of copper-chloride crystallization and silver steigbild images, and 'Gärprobe', a measure of how good the milk is for cheese-making. Her

concerned with an object but with the whole organism and with a process which maintains health. By considering these things we might even find a medicine which works. The word 'Biodynamic' comes from the two Greek words bios, life, and dynamis, energy. Let's compare this with the modern term 'antibiotic' as in 'antibiotic medicine,' which word means, literally, anti-life: anti-life medicine! The issues here involved are fairly fundamental.

Modern cows, fed with steroids, milked by machine and deprived of horns, don't even get sex. Instead, the semen is thawed out from being cryogenically frozen in liquid nitrogen and then just injected. Farmers select the genes of the bull of their choice. What kind of view of Creation is this, to suppose that the dumb creatures can be so treated without repercussions on us? The public are now reacting against the dictates of a materialistic science, because it keeps throwing up disasters as Mother Nature reacts against it. Biodynamic farmers received a boost when a prophetic statement made in 1923 by Rudolf Steiner concerning BSE was unearthed: if cows were fed on meat produce, he explained, they would go mad.[43] And so they did.[44]

This work is offered as a contribution to the theory and practice of organic gardening and farming. In practical terms, it may well contain nothing as important as how to prepare a good compost-heap. But, use of optimal sowing times as here described should be able to increase both quality and quantity of products, but by how much is a question that should presently remain open-ended. If compelled to give an arm-waving figure, I'd surmise we are talking about something between ten and twenty percent. But also, tuning in to the cosmic cycles helps to reduce stress levels, and gives more of a sense of purpose and dignity in life, which farmers could well use today, by experiencing themselves as working in accord with the deep rhythms and patterns of Mother Earth.

PhD is at Kassel University. 'Milk Quality of horned and de-horned cows', Star and Furrow Winter, 2007, pp.10-12.
43 Rudolf Steiner, Lectures to workmen at the Goetheanum, 13th Jan '23; reprinted as, *From Comets to cocaine, Answers to questions*, 2000, Rudolf Steiner Press, p.228. The madness would be due to uric acid production, as would go to the brain.
44 Let's have a confirmatory statement by British Biodynamic (wine) expert Monty Waldin: 'No certified Biodynamic cow has ever suffered from mad cow disease or has had her horns removed or has been force fed her own meat.' (The Ecologist, 1.3.09)

2. Four Kinds of Month

Compelled by fates that likewise rule the sea / I roll out month-long periods of time / In sure-returning cycles. As the light / Of glorious beauty slowly leaves my face / So does the ocean, flowing from the shore / Lose its increase of waters in the deep.
Riddle by Aldheim for King Aldfrith[1]

We turn next to the main astronomical concepts involved, in terms of our experience of the night sky. Not the least value of using a Biodynamic calendar is that it encourages an experiential attitude towards these things. The gardener is personally enriched by being able to see at night the situation that has been used in the daytime with the planting calendar - in contrast with the poor city dweller, surrounded by night's neon glow.

Of the two astronomical chapters in this work, the present one concerns the different kinds of months that are involved. A later chapter ('The Starry Script') concerns the constellations of the zodiac, and it seemed feasible to leave that more abstruse issue until after the main bulk of evidence had been presented. It would be possible to skip the present chapter and just return to it as needed, however it is preferable if one can understand the basic frameworks, which can otherwise generate so much confusion.

Astronomically, there are four basic lunar months. These are:

The Synodic or phase cycle	29.5 days
The Nodal cycle	27.6 days
The Sidereal cycle	27.3 days
Apogee-Perigee cycle	27.2 days

These four cycles of the Moon can be viewed as relationships: its meeting with the Sun (phase or synodic), its approach to the Earth (apogee-perigee), its touching the plane of the Earth's orbit (nodal) and its motion against the stars (sidereal).[2] Contrary to widespread and persistent belief, there is no such thing as a 28-day lunar month.

---

1 Quoted in Tides and the Pull of the Moon by F.E.Wylie (1980, Berkeley Books, New York p.41) who obtained it from 'Northumbria in the Days of Bede' St Martin's Press NY 1976. Aldeheim, Bishop of Sherborne, lived c. 640-709. Wylie's book is recommended for its account of marine tides, and 'the tides of life.'
2 Otto Neugebauer, A History of Ancient Mathematical Astronomy, Part 1 (Springer-Verlag, NY) 1975, p.310, describes how these four monthly cycles were known to the ancient Chaldeans.

Biodynamic calendars nowadays incorporate five different monthly lunar cycles,[3] which makes them very complicated, too much so in my view. They use what they call the ascending-descending path of the Moon, which is based upon the tropical month period. This is of virtually the same duration as the sidereal month, but conceptually it is quite different (see below). Our 'Planting by the Moon' did not view this as a primary monthly cycle. Its use by farmers has its roots in folk tradition.

The Synodic Month

The word synodic means meeting, as in a 'synod' of bishops. The two luminaries meet in the sky every 29.5 days, on average. The period is shorter by several hours in winter than in summer. In its changing phases, the Moon mirrors the Sun's light from different angles, vanishing from view several days every month at its new position. In ancient Rome, the 'Ides' meant the time of the Full Moon, as in the 'Ides of March', and formed the middle of each month, when religious ceremonies were held.[4]

There are twelve and one-third such cycles in a year. When calendars used to be lunar, an extra, thirteenth month had to be 'intercalated' every third year. This still happens with the Jewish and Hindu religious calendars, which are lunar-based.[5] In contrast, the Muslim religious calendar is lunar but has only twelve months, so that its months including Ramadan have to move round a third of a month each year, against the secular (solar) calendar. The month of fasting in this calendar begins and ends by the first sighting of a New Moon. The root meaning of the word 'calendar' is from 'calare' meaning to cry out, which is what the priest used to do at the commencement of the month, as the first thin sliver of the New Moon appeared in the morning sky.

The modern age has left these things behind, so that diaries no longer even contain the lunar phases. I personally take quite seriously the suggestion made by Dr Leiber in his book, The Lunar Effect, that: "Ancient civilizations used lunar calendars; the Israelis, the Chinese, the Hindus and the Moslems still do. Interestingly, there is much less violent crime in these societies".[6]

---

3 Maria Thun, 'Work on the Land and the Constellations', 1977.
4 A solar calendar was first introduced by Julius Caesar: when he was stabbed on the Ides of March, this was no longer a lunar-determined date, it was just March 15th.
5 These sacred calendars used the Metonic cycle, to keep months and years in step. Over a nineteen year interval, seven long years are counted, each having thirteen lunar months, plus twelve short years which have only twelve lunations.
6 Dr Arno Lieber, The Lunar Effect 1979, p.136.

Four Kinds of Month

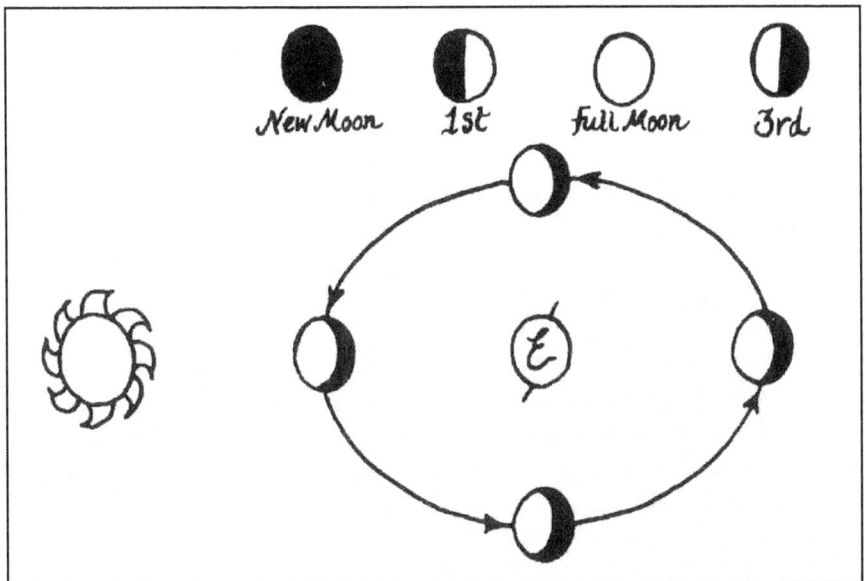

Fig 2.1 The Four Phases of the Moon

At the start of the 1980s, two different yearly lunar calendars were available for gardeners in the UK, and both used four lunar-month cycles: 'Planting by the Moon', and the English translation of the Thun calendar. A major difference between the two was that the former used the phase or synodic cycle while the latter did not. There is something of a paradox here, as will unfold in due course: when the subject of lunar gardening is raised, people tend to recall some adage about planting peas or whatever before the Full Moon; and yet, this cycle was for long absent from the main Biodynamic calendars. Millennia of tradition cluster around the connection of this fundamental cycle with the concepts of fertility and growth, and we will see how laboratory experiments are to some degree supporting such a linkage.

There is more evidence linking this monthly cycle to plant growth than any other. It appears to work through the water-element, through fluid processes, reminding one perhaps of the 'riddle' cited at the start of this chapter. The riddle there presented is that a monthly cycle in the heavens generates a tidal rhythm of half that length, i.e. 14-days.

This paradox remained opaque to human reason until the year 1685, when Newton resolved it by pondering the joint movement of the Earth and Moon around their common centre of gravity. When looking at the synodic rhythm in a biological/ecological context, one commonly finds either a monthly or a fortnightly rhythm present in a given time-study, and one cannot predict which of these will appear. We will here look at studies showing how the

Moon's global effect upon geomagnetism is a monthly oscillation,[7] whereas that upon rainfall is fortnightly - peaking, rather like the tides, just after the Full and New Moons.

The Greek word 'synodos' means meeting, but also signifies 'copulation', as is relevant to the linkage of this cycle to fertility, conception and birth. There was as we saw a curiously exact multiple of nine meetings of the two luminaries over the mean interval between conception and birth. There is a certain analogy between the disintegration of the womb lining with the growth of a new one over four or five days, and the 'dying' of the old moon with the appearance of a new one over a similar period. One appreciates that educated persons will tend to dismiss this as mere primitive logic.

Newton's writings often contained the word 'menstrual,' but his use of this term was solely astronomical, pertaining merely to the synodic monthly cycle, and contains no hint of any physiological context. The meaning of the word has shifted in the course of three centuries, from astronomy to physiology. The discreet French phrase "le moment de la lune" for a woman's period echoes this old meaning. We may think of this cycle as the blood-rhythm.

The Pull of Perigee

The Moon draws nearest to us at perigee and recedes furthest at apogee each month, a change in distance of fourteen percent. This monthly cycle affects the pull of the tides. Flood tides are more likely when Full Moon aligns with perigee, as tides rise thirty percent higher at perigee than apogee.

It was argued by the Florida psychiatrist Dr Lieber that the strong pull of perigee aligning with the Full Moon tended to produce more crimes of violence and homicides.[8] This is the time when floods are likely at the spring tides, because of the extra strong gravity pull, and Lieber argued that there was some sort of analogy with the dire Hecate-type influence he claimed to discern.[9] It must be said that few found his statistics convincing. A relevant quote here comes from Shakespeare's play Othello. When Othello has strangled his love Desdemona, he at once blames the Moon:

> It is the very error of the Moon,
> She comes more near the Earth than she is wont,
> And makes men mad.

---

7 B.Bell and R.Defouw, 'Concerning a lunar modulation of Geomagnetic Activity', *Journal of Geophysical Research*, 1964, 69, 3169-3174.
8 Lieber ref (6), Ch.2.
9 The analogy with the tides that Lieber was trying to establish, had the problem that spring tides reach equal maxima at the Full and New positions, and so are at double the frequency from an effect that peaks only monthly, at Full Moon.

-a Shakespearean pun on the latin 'errare', to wander. Some studies have shown stress-related conditions to be related to the perigee position.[10],[11] In general however there are no astrological traditions pertaining to perigee - in contrast with the nodes, it lacks a traditional significance.

The apogee-perigee cycle is the only one here considered that is not 'seen' in the night sky. It can however be discerned in a lunar calendar, by the zodiac ingresses: these appear as shortest in duration around the position of perigee, where a single sign lasts for merely two days, due to the Moon moving faster as it draws nearer to the Earth, while at apogee the converse applies with the signs lasting two-and-a-half days.[12] Biodynamic calendars caution to avoid sowing at the perigee position, for around twelve hours on either side, believing that some perturbing influence is then likely to manifest.[13]

The perigee rocks erratically back and forth in its nine-year sojourn around the zodiac. Twice a year it turns to move retrograde, and then can regress by more than a whole zodiac sign between one perigee and the next, after which it moves forward again more slowly.[14] If perigee is to be included in a gardening calendar, locating its positions for the forthcoming year is dependent on a computer program, whose results are printed in astronomical tables. The perigee will stray far from a notional 'apse line', which represents a mathematical average of these erratic motions. The apogee position, as the furthest point from Earth, has a less energetic motion, and its retrograde motion covers only a few degrees. The apogee and perigee positions do not oscillate in synchrony. Astronomers talk about

---

10 Lieber (ref.6, p.51) reported a study of eleven thousand cases of aggravated assault taken from Dade County Public Safety Department in Florida 1969-73. They showed a large peak at Full Moon.

11 Simon Best, 'Stressful Moon' The Astrological Journal 1981: Part I Winter p.37, Part II Spring p.78; also his 'Doctor' article, 1980, July 24 p.35. Dr Lieber's evidence for the anomalistic period was little more than suggestive, with a peak in calls made to the police at apogee.

12 For mathematically-inclined readers, the ratio of lunar angular velocities apogee/perigee is the square of that between their distances, owing to Kepler's second law: the apogee distance being 1.14 that of perigee, the ratio of rates of movement will be the square of this, ie 1.30; thus the Moon moves 30% faster at perigee than at apogee.

13 Sattler & Wistinghausen, 'Biodynamic Farming Practice', CUP, translated from the German 1989, p.101, reported concerning the anomalistic cycle: 'Heinrich Schmid, a farmer in Korbach, Germany, and Franz Rulni have been investigating the connection between moon rhythms and the mating of cattle. Their findings suggest that apogee brings out more the male type, perigee the female.' A letter of inquiry as regards the data elicited no reply. Franz Rulni's original Biodynamic calendar (Ch. 6), made such claims of gender prediction, which have dropped out from more recent calendars.

14 Joachim Schultz, *Movement and Rhythm of the Stars* 1963, translated from the German by John Meeks, Floris Books 1986, 2008, p.89.

the 'apse line' as linking mean apogee and perigee positions, which revolves once every nine years round the zodiac. The line is a bit of an abstraction, since its two ends move in this irregular manner.

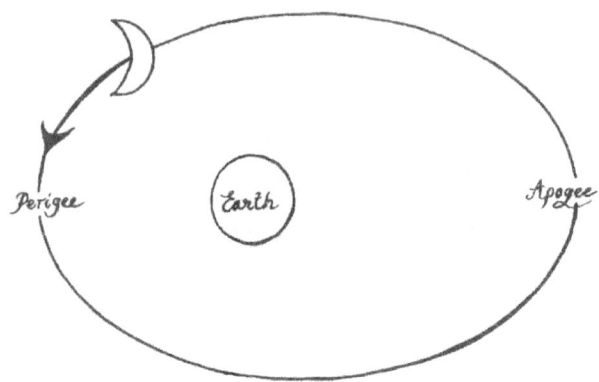

Fig 2.2   The Moon's apogee-perigee cycle, of 27 days

At perigee each month, 'moonquakes' maximise. The Apollo missions of years gone by left apparatus on the lunar surface to record them. Luna's structure is highly resonant, as was found by impacting a satellite onto its surface, when its echoes continued to ring for three hours afterwards. Moonquakes register as a light, tinkling sound, caused by gravity stresses, and they rise to a crescendo as the Moon draws nearest to its parent planet each month.

The closest perigee of the year is called 'proxigee', and is associated with floods and more extreme weather conditions. This normally falls on the day of a Full Moon, but sometimes on a New Moon. Of late, people have been alluding to these extra-strong Full Moons as 'supermoons.'[15]

---

[15] These extra-strong moons happen every 413 days, because 14 lunar months = 413.4 days and 15 anomalistic months = 413.3 days, i.e. the two monthly periods then coincide. This intriguing coincidence puts the closest moon of the year at perigee and at Full Moon (over 2008 to 2020: it re-sets now and then.)

The Nodal Cycle

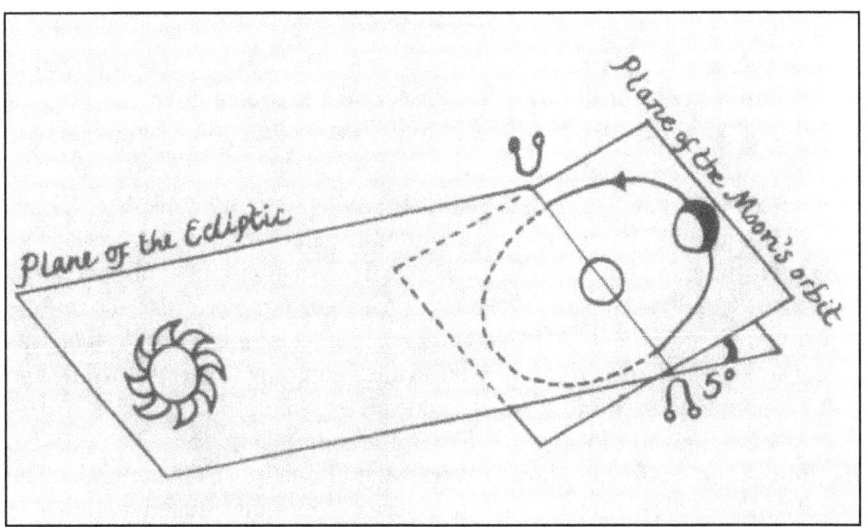

Fig 2.3 The plane of the lunar orbit, showing the lunar nodes.

Traditionally, the two node positions were called 'Dragon's head' and 'Dragon's tail', as if some menacing dragon were curled around the zodiac, liable to swallow up the Sun at an unpredictable moment. The nodes are the points where eclipses occur, on the line of intersection of the ecliptic with the plane of Luna's orbit.

The ecliptic is the plane in which the Earth orbits the Sun, which we may picture in the night sky as a line across the heavens, along which the planets wander. If the reader has not done it before, it is a worthwhile exercise to locate that line. Choose an evening after the Sun has just set so its point of setting can still be located, when two other objects are visible on the ecliptic, e.g. the Moon and the bright planet Jupiter. One can then discern the path of the ecliptic, passing through these three.

Ah, but later in the night, can you still see the ecliptic when the stars are out? One should try to recognize a constellation or two which mark its path. Watch as the Moon passes by some bright planet, e.g. Jupiter. Each day it moves thirteen degrees, and through successive nights one can discern that span. When it passes by, does it pass to one side of the planet, or are the two close together?

This matter hinges upon the nodal cycle, for Luna can move almost six degrees away from the ecliptic, as it swings from side to side each month, winding its serpent path from one side to the other.

The nodal cycle causes large, monthly effects on rainfall and on the Earth's magnetic field. Studies have shown that geomagnetic activity peaks at days of Full Moon, this effect being very marked when these occur on the ecliptic (i.e. when a Full Moon occurs at the node position),[16],[17] when it increases by as much as 30%. This effect diminishes then finally disappears altogether away from the node positions, when the Moon is at greater than four degrees of celestial latitude.[18] The Moon's latitude, as opposed to its longitude, can be up to five or six degrees.

Large-scale surveys, in both the northern and southern hemispheres, have shown that rainfall patterns show a bimonthly, tidal rhythm, peaking a few days after the Full and New Moons.[19],[20],[21] This fluctuation, of about twenty percent of the mean rainfall, recalls Shakespeare's words about "the governess of floods" in A Midsummer Night's Dream.[22] This effect was found to be modulated by the nodal cycle, being maximal when Full and New Moons occurred close to the nodes, i.e. at celestial latitudes of less than one degree, and far weaker for larger zodiac latitudes.[23] For rainfall, the Full Moon's effect is more powerful the closer it is to the ecliptic. The nodal and synodic cycles interact in patterns of rainfall frequency, rather as we saw the apogee-perigee and synodic interacting in tidal patterns. Both involve two lunar-monthly cycles. There is here a contrast between the rhythms of rainfall and the tides. The latter depends upon gravity, and is

---

16 Op. cit. (57), also B.Bell and R.Defouw, 'Dependence of the lunar Modulation of Geomagnetic Activity on the Celestial Latitude of the Moon,' *Journal of Geophysical Research*, 1966, 71, 3, 951-957.
17 Stolov and Cameron, 'Variations of Geomagnetic Activity with lunar phase,' Journal of Geophysical Research, 1964, 69, 4975-4982. For a review, see Herman & Golding 1978, 215-221.
18 For a fine account of how the lunar monthly effect upon the GMF is modulated by lunar celestial latitude, i.e. the lunar node cycle (reviewing the work of Bell, Defouw, Stolov and Cameron), see: J Herman and R Goldberg, 'Sun, weather and Climate', 1978 NASA, pp. 217-220: '...there does appear to be a lunar modulation effect on geomagnetic activity when the Moon is within 4° of the ecliptic plane on the morning side of the Earth (i.e., between full Moon and last quarter positions)', p.220.
19 Bradley, Woodbury and Brier, 'Lunar Synodical Period and Widespread Precipitation,' *Science*, 1962, 137, pp.748-9.
20 Brier and Bradley, 'Lunar Synodic Precipitation in the United States,' *Journal of Atmospheric Sciences*, 1964, 21, 386-395. The rainfall peaks fell three days after the Full & New Moon positions, and were considerably more pronounced for years of low solar activity in the 11-year sunspot cycle.
21 E.Adderley & E.Brown, 'Lunar Component in Precipitation data' *Science*, 1962, 137 749-750. For an account of the simultaneous publication of these two lunar-month rainfall studies, one in Australia and the other in the US, see Lyell Watson, *Supernature*, 1973, p.27-8.
22 Shakespeare, 'A Midsummer Night's Dream' : 'Therefore the Moon, the governess of floods,/Pale in her anger, washes all the air,/That rheumatic diseases do abound.'
22 Bell and Defouw op. cit. (15), 1966.

related to the apogee-perigee cycle. The condensation of water vapor into rain is a more electrical matter, and so it should not surprise us that it follows the same cycle as does the earth's geomagnetism, fluctuating in tune with the nodal cycle.

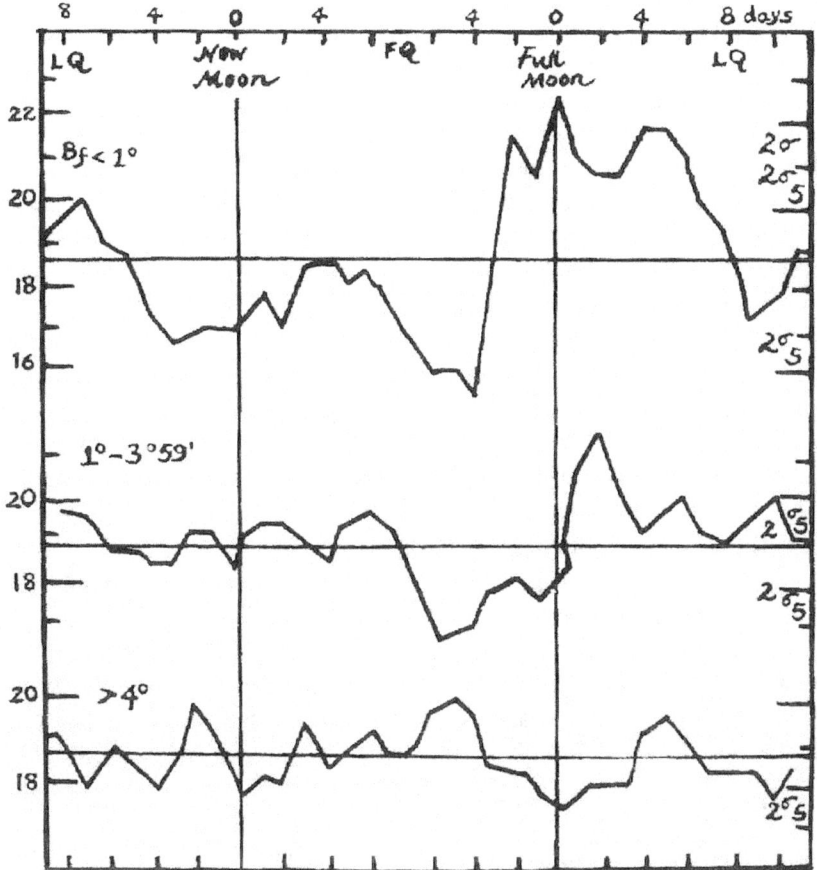

Fig 2.4 From Bell and Defouw (1964), showing how the perturbations in the GMF (Geo-magnetic field) over the lunar months varied according to celestial latitude (ie the node cycle), being maximal when syzygy is aligned to the nodal axis. The top graph shows the lunar month with GMF strongly peaking around Full Moons that were within one degree of celestial latitude from the ecliptic, while the bottom graph averages lunar months whose F.M.s were four degrees or more away from the ecliptic.

Thus the traditional image of some dragon-energy linked to the nodes may have something in it. Solar and lunar eclipses occur when the nodes come into conjunction with the Full and New Moons respectively. But, as well as that, electrical and magnetic perturbations in the biosphere are as we have seen associated with the nodal cycle, via its effect upon the synodic cycle.

Luna orbits in its plane tilted at 5° to the ecliptic, so that twice monthly it cuts the ecliptic plane at the nodes. The line joining the nodes is the nodal axis, and this revolves once per 18.6 years around the circle of the zodiac. Its motion defines the eclipse seasons, six months apart, e.g. June and December. These are the months of the year wherein eclipses cluster, and take eighteen years and seven months to revolve once round the calendar, reflecting the motion in space of the nodal axis. Astrologers associate nodes with the notion of fate, and in Indian astrology the South node represents one's past karma while the North node indicates future prospects. Biodynamic calendar make no distinction between the two: as plants are a more rudimentary form of life, one applies these events in a less differentiated manner.

If Luna were a separate planet on its own, then its orbit ought to lie in the plane of the ecliptic, while if it were properly a satellite of the Earth, it should be near to the Earth's equatorial plane, which is tilted at twenty-three degrees to the ecliptic. In the latter case there would be no doubt where it had come from, namely Earth.

Luna asserts its independence by having its own plane of orbit, neither in the ecliptic nor on Earth's equator. Its orbital plane is unique, and astronomers cannot find any reason as to why it should lie therein. The angle of tilt conveniently gives us eclipses only now and then and not at every New Moon. The syzygy axis, the line linking the Full and New positions, has to align with the nodal axis to produce an eclipse. Twice a year they swing into such alignment, giving the above-mentioned 'eclipse seasons'.

The Sidereal cycle

The sidereal orbit of the Moon (pronounced as in 'side') is of central importance for the present work. Britain's Biodynamic journal is entitled 'Star and Furrow', which would seem to proclaim fairly definitely that the sidereal cycle has some agricultural relevance.

Or, on the other hand, has it all been a huge mistake, as a weighty opus by Dr Spiess has argued,[24] this two-volume work being published by the Forschungsring of the German Biodynamic farmers? Later on, we grapple with this issue.

Luna revolves on its own axis in space once per 27.3 days, facing ever earthwards. Curiously, we may note that sunspots vary in their rotation rates depending on their latitude in the Sun's surface (the sun being a gaseous

---

24 Hartmut Spiess, 1994 'Chronobiologische Untersuchungen...', Darmstadt; 2 vols.

body and rotating differentially) but the average figure for their rotation rate that scientists use is widely taken as being also 27.3 days.

That is as seen from Earth, i.e. it takes that long on average for a cluster of sunspots to rotate around the Sun and return to the same position. Finally, the diameter of Luna is 27.3% that of Earth. This involves mean diameters, as they vary a little from pole to equator. So, this number turns up curiously in different contexts, as well as being the fundamental orbit period.[25]

To help get a feel for this, let's consider an early - nay, the earliest - expression of the two fundamental lunar cycles. The beginning of astronomy may well have been aided or stimulated by the practice of agriculture. Long ago, in British pre-history, the huge stones of Avebury were brought together, to create Europe's largest megalithic monument. There is now a village at its centre, where once stood two stone circles: one of these had twenty-nine stones and the other, twenty-seven. The gaunt remains of these two awesome stone circles can still be seen next to each other, like interacting gear wheels. They are of similar size, a hundred meters in diameter, and stand at the very centre of the giant Avebury complex.[26] It is inspiring to realize that early Britons wanted to focus on these two primary lunar-month cycles.

Books on Avebury almost invariably omit or misrepresent these numbers, and make no comment on their significance, the topic being a deeply forgotten aspect of our culture: the current image of 'primitive man' does not well accord with megalithic Britain's primary lunar temple and foremost cultural centre having enshrined these two fundamental lunar-month periods. These two months represent different kinds of time.

Avebury had a deep connection with Stonehenge which is just a few miles away and exactly due South, and we can best experience this by considering the outermost ring of Stonehenge which is made of 'Aubrey holes.' There are fifty-six of these holes, which is the sum of twenty-seven plus twenty-nine. The outer ring of holes at Stonehenge is the arithmetic sum of the two inner stone rings at Avebury - as well as being the same size, a hundred meters across. But, whereas Avebury was purely lunar, Stonehenge was solar-lunar, its function was more to balance and integrate the solar and lunar principles. Stonehenge's design alludes to the 18.6 - year nodal cycle, relevant to eclipse prediction and, as we will see later, quite fundamental for long-term agriculture planning, but it doesn't have any 'sidereal' connection as did Avebury: it wasn't connected to the realm of the stars. As the great stones were smashed up, which happened mainly in the early

---

[25] For mathematically-inclined readers, this is related to the 29.5 -day synodic lunar month by the equation: $1/29.5 = 1/27.3 - 1/365$
[26] Caroline Malone, *The Prehistoric Monuments of Avebury* 1990 English Heritage, p.11; Evelyn Francis, *Avebury*, Wooden Books, 2000.

eighteenth-century, the 'starry wheel' of 27 stones at Avebury was destroyed more completely than any other, with only a few left standing, and one can only see it now in old pictures.

In the seventeenth century, this 27-day sidereal month was used by Isaac Newton to demonstrate his gravity theory. His maths linked together the fall of an apple and the orbit of the Moon. In the firm words of Gardener's Weekly, it "caused him to discovery gravity".[27] Thereby Luna's sidereal motion became explained in terms of a pull towards the centre, i.e., earthwards. Before this, its path had been experienced in terms of the star-constellations beyond it, out in the cosmic periphery. Thus Isaac Newton, in his awesome gravity computation performed in the year 1685, linked Earth's gravity field to Luna's revolution against the stars.

Can the constellations really have an effect on Earth, if the nearest star is light-years away? It may be helpful to make an analogy here with physics: a Foucault pendulum is a very large pendulum that will keep swinging for days. A Frenchman called Foucault discovered in 1851 that such a pendulum maintained its orientation in accord with the position of the distant stars.

He demonstrated it using a huge pendulum in the Paris Pantheon, where the plane in which it swung gradually revolved in the course of a day. Were such a pendulum set up at the North Pole, then the plane of its oscillation would revolve once per sidereal day, that is 23 hours and 56 minutes. In that period the earth revolves once against the stars. The pendulum keeps itself aligned with the stars.

To quote from a physics textbook, concerning the 'Foucault pendulum' as it is called:

> Experiments in mechanics show that there exists a reference frame that is identical to that defined by the position of the stars.[28]

Somehow, what a physicist calls an inertial reference frame is linked to the position of distant stars.

The Tropical Month

The word Tropos means 'turning'. At the tropics of Cancer and Capricorn the Sun was experienced as turning round in its yearly course. Thereby it reaches the sign of Cancer at northern midsummer and Capricorn at midwinter. The Moon reaches those same points once per tropical month. This period is hardly distinguishable from the sidereal month, only differing

---

27 *Let Newton Be!* Ed Fauvel et al, OUP 1988, p.235.
28 Ray Skinner, *Mechanics*, 1969, p129.

in the fourth place of decimals, but conceptually it is different because it is not related to the stars. The periods are:

Sidereal month - 27.3216 days

Tropical month - 27.3212 days

The tropical month pertains to the period for which the Moon is visible above the horizon, ranging from nine to fifteen hours. This effect varies greatly with latitude: in the North of Scotland, the Moon in the zodiac sign of Cancer will only dip below the horizon for a short period each day. In American almanacs, the Moon is said to be 'riding high' when it describes such large, high arcs in the night sky, and 'riding low' on nights when it remains low on the horizon. Following the example of Dr Spiess, we will be referring to this cycle as the tropical month, as its proper astronomical name.

Biodynamic calendars have transferred to this cycle all the gardening tasks that tradition had assigned to the phase cycle, such as pruning and transplanting.[29],[30] How this came about is not easy to discern. Up to the 1980s, the Thun calendar embodied four cycles: the sidereal, tropical, nodal and apogee-perigee, together with some planetary aspects. In 1982 the phase cycle was added, not because it had relevance for farming or gardening, but because many readers had been confusing the tropical cycle with it.

Kepler, in the Preface to his 1602 Calendar affirmed that the Moon's influence was greater on days when it rose highest in the sky.[31] In the 1950s, an early Biodynamic calendar by Franz Rulni started using this tropical cycle, apparently picking up on a folk-tradition from the Emmenthal valley in Switzerland. He used the terms 'nidsi' and 'obsi' for two halves of this cycle, and today BD calendars use the equivalent terms 'ascending' and 'descending'. During northern spring the Sun is moving from Capricorn towards Cancer. (These are the 'signs' as astrologers use the term and not

---

29 Maria Thun, *Working with the Stars:* 'When the daily arcs of the moon become even lower we speak of the moon 'descending'. This period we always designate as planting time.' Planting here alludes to the moving of young plants from their seedbeds to their final positions.

30 Elizabeth Vreede, 'The Ascending Periods of the Planets at their times of Special Influence' Anthroposophical Agricultural Foundation *Notes & Correspondence,* 1936, IV, 345-350 (translated from the German). Vreede discerned 'a strong upward striving and lengthening tendency' in plants over the ascending period of the (tropical) lunar month, and a weakening earthward inclining tendency during the descending period; just as Saturn's 15-years in ascending phase of its cycle was viewed as the time to plant Saturnine trees such as beech.

31 Kepler, *De Fundamentis Astrologiae Certioribus* 1602, Section XXXIII, Field 1984, p.248: 'For both the planets and the Moon operate most strongly from Cancer, because in that sign they are longest above the horizon...'

the constellations). Likewise, spring-type energies are discerned by northern Biodynamic farmers as the Moon is 'ascending', ie moving from Capricorn to Cancer.

If the above is confusing, the following may, just possibly, be of assistance. In midwinter in the northern hemisphere, the Full Moons arc up near to the zenith of the heavens, while those of midsummer remain low on the horizon. This is because in midwinter, the Sun is in Capricorn and therefore the Full Moon, being diametrically opposite in the sky, is in Cancer. Cancer is the Sun's position at midsummer, when it rises highest. These (tropical) signs are firmly anchored to the seasons of the year: the Sun at zero Capricorn is by definition at the winter solstice.[32]

I am doubtful as to whether Biodynamic farmers should be using this cycle.[33] If Biodynamic farmers know of some basis for using it in their calendars - with a 'descending' moon in Australia while it is 'ascending' in the Northern hemisphere - then it is up to them to tell us. The most that can be said for it is that it probably has quite a good pedigree in folklore tradition, having been referred to by Kepler and now found in gardening calendars on both sides of the Atlantic, though in different forms.

A Biune System

Astronomers have a headache accounting for why a sphere one-quarter of Earth's diameter should have ended up so far away - thirty Earth-diameters - in a plane all of its own, and yet have a nearly circular orbit. Were it a little nearer, or more eccentric in its orbit, then countries would be flooded periodically by giant tides, but fortunately this is not so. We may sense a just-so balance in these matters.

The Moon's distance determines the all-important ratio of twelve months to each year: were it a mere few percent nearer to the Earth, there would be thirteen months in a year, and no four seasons or a zodiac divided into twelve. The Chaldeans divided the circle into 360° as twelve sectors of thirty degree divisions, reflecting a schematic calendar system of twelve months of thirty days, with the Sun moving one degree per day. It is the nature of these primordial cycles, which formed the matrix for life's development, that we can hardly imagine them as other than what they are.

It is worth pondering the mysterious relationship between the two spheres in space. Perhaps this can tell us something about the mystery of life's

---

32 Altitude is height above the horizon, while declination is angular distance from the celestial equator. The 'ascending' half of the tropical month is its period of increasing declination.

33 Herbert Koepf in *The Biodynamic Farm,* 1989 wrote, concerning beliefs in alleged effects of the tropical month (p.115): 'But these are traditional views, which at this point are not really backed by recent observations or experiments.'

origin. One sphere appears as focussed upon the other. The huge craters and the long ray formations across its surface that give Luna its mysterious appearance are all on one side, that facing Earthwards, as are its huge lava seas with high concentrations of strange heavy metals such as titanium and uranium. The largest craters cluster on a line down its central meridian, facing towards us. The far side has little more than just mountain ranges, and is quite uninteresting.

The two spheres revolve around each other, opposites in every respect: one is bone-dry while the other is an emerald watery sphere; one devoid of life while the other teems with it; one ravaged by huge craters while the other shows hardly a trace; one showing extreme old age in its rocks while the other is a fountain of youth by comparison; one with a powerful magnetic field and the other with hardly any; and one with an independent axial rotation while the other is phase-locked into facing its parent planet. The satellite is more primitive in its elemental composition, apparently formed in some early period when there were fewer elements about, or as if it had passed through some high-temperature process which had removed water and low-vapourisation metals such as sodium, potassium and calcium. The surface moondust had pretty glass spherules of all different colors... Theories formed by the selenologists over its origin never made much headway against the baffling data carried back by the Apollo missions.[34] I surmise that they were insufficiently bizarre to get a grip on the phenomena.

Luna is believed to have caused Earth to develop a magnetic field strong enough for life to develop. The geomagnetic field protects life from deadly radiation coming in from outer space. The Moon stirred up Earth's core over the aeons by its pull, thereby establishing the geomagnetic field.[35] The adjacent planets Mars and Venus have nothing resembling this field, lacking such a satellite to have done this.

Such considerations are of value in this hurried age, for the enigmas involved are timeless. Pondering them can help one to achieve inner calm. We cannot hope to fathom them, yet it is beneficial to dwell upon such things from time to time. No-one can comprehend how Earth and its companion came to be as it is, or how it affects the human psyche, and the germination of seeds. To quote Frederick Nietzsche, 'This world is more deeply wrought than meets the eye of passing day.' Scientists measure its distance from us to a centimeter with laser beams, but their inorganic approach has forgotten the all-important thing, its effect upon life-processes. A dry and lifeless orb modulates water- and life- processes in a plant; that is the mystery.

---

34 Zdenek Kopal, Man and His Universe, 1972, Ch.VII.
35 Hoimar von Ditfurth, *Children of the Universe, The Tale of Our Existence* translated from the German, 1975, p.175.

3. Lunar Phase Rhythms in Plant Growth

> *"In the plant world in temperate latitudes, it is immediately obvious that the germination, growth, maturation and perennial structure formation in trees are marked by an alternation between active and resting phases."* - Ernst Zürcher 'Lunar rhythms in Forestry Traditions, 2001, p.463.

Rhythms of activity in plant growth follow the changing Sun-Moon angle. Laboratory and field studies of 'the solunar clocks of life' have shown this in plants, to use the fine phrase coined by the late Frank Brown, biology professor of Northwestern University, Illinois. Such studies may lack an immediate relevance to the practical gardener concerned with growing crops, but they form an essential background to our story. To some extent one can divide our subject into pure and applied branches, and this chapter is more concerned with its pure or theoretical side.

The subject touches upon the fascinating issue of rhythms in living organisms, and the manner in which they are timed. Humans and animals have internal rhythmic processes, e.g. the heart-beat, whereas plants hardly have any such, their rhythmic processes being timed externally. Biochemists may continue to seek for a timing 'mechanism' within the plant, but it seems increasingly likely that this hardly exists in the form in which they are looking for it, and that the rhythms to which plants respond are, to use the technical term, 'exogenous' or externally generated.

Plants are inherently adapted to the cycles of the day, month and year. That is to say, removed from their natural environment and grown in a laboratory, under constant ambient conditions, plants continue to manifest growth-rhythms of these fundamental periods. A classic example is the experiment of the German professor Bunning, a German physiologist at the University of Tubingen, which looked at seed germination through the course of the year. Bunning kept seeds under uniform conditions in his laboratory and found that they germinated optimally in the springtime and least well in late autumn. Light and temperature were held steady and so could not account for this. Another batch held at a different temperature still showed the same yearly cycle in fertility.[1] The conclusion to be drawn from this experiment is far-reaching, namely that when in the springtime things start to grow and burgeon, this is not simply because temperature and daylight have

---

[1] E. Bunning, 'Endogenous Rhythms in Plants', Annual Review of Plant Physiology, 1956, 7, p.86; see also, Michel Gauquelin, The Cosmic Clocks 1973, p.114-5.

increased. Rephrasing that in more positive terms is difficult, as it involves a concept that is so simple as to be difficult to put into words: that there exists something which we might call the Fertility of the Earth, which is maximal at springtime.

Farmers have always known this, you may say, which is true enough. But the science of biology does not acknowledge that living things are able to respond directly to such a cycle, holding that only external factors such as temperature can account for such a response.

Bunning's experiment demonstrated that there was no clock inside the seeds measuring the cycle of the year, because that would have been affected by temperature - normally, biochemical reactions are assumed to double in rate for each ten degree rise in temperature. Walking down one's garden path, observing how plants of the same species burst into bloom within a few days of each other, one can reflect on this time-experiment, admirable for the simplicity of its conception. I tend to call it the Persephone experiment.

The science of chronobiology finds these three time-cycles, day, month and year, embedded in the time-structure of living organisms. How these are to be 'explained' need not greatly concern us. The one scientist who seemed really to understood 'the clocks of life', was the late Professor Frank Brown, and we refer to his work in this chapter.[2] His colleagues never liked his approach, because it did not start by cutting open small animals, but instead pointed to what we might call a more holistic view.

Biologists note the adaptation of flowers to the course of the year, whereby they bloom once a certain day-length is reached, but the course of the month has dropped out of their equations. From one point of view this is understandable, the light of the Moon being three hundred thousand times dimmer than that of the Sun. One's common-sense might tell one that any influences would be proportionally weaker: one's common-sense would here be mistaken.

To take an example from animal behavior, sea-horses around Florida produce eggs at the Full Moon. In a biology department of Cambridge University, these creatures were kept in an aquarium under scrutiny, and continued to lay their eggs at the Full-Moon.[3]

---

2 F.A.Brown, 'A Hypothesis for extrinsic timing of Circadian Rhythms' *Canadian Journal of Botany,* 1969, 47, p.287; see, e.g., Cloudesley-Thompson's *Biological Clocks,* 1980; 'Biological Clocks and the role of subtle geophysical factors', H.Marguerite Webb, in *Geo-Cosmic Relations* Ed. Tomassen, Pudoc, Netherlands 1990 pp.56-64; and, in the same volume, B.G. Cummins, 'Biological Cyclicity in relation to some astronomical parameters - a review', pp.31-56.
3 Work by Dr Amanda Vincent at Cambridge Zoology Department (unpublished).

Does this mean that the sea-horses have a calendar somewhere? If so, the scientists have been unable to find it. Can the sea-horses perhaps sense lunar phase as a time-signal? The question recalls the theory we looked at earlier, that womankind was once partially or fully synchronized in its monthly cycle, with ovulation linked to the Full Moon.

Do humans respond to the lunar cycle? Endless inconclusive papers have been published on this question, and centuries-old debates on the matter are today no further advanced than when they begun. If anything they are less well resolved, as the early papers on, for example, human birthrate and the lunar cycle seemed to show such a linkage quite well, whereas modern urban living conditions have fragmented if not wholly erased the phenomenon.

One definite fact that psychologists have established, in the well-known 'bunker experiment,' is that persons isolated from daily influences and away from sunlight for a period came to set their routines to a 25-hour lunar-day rhythm rather than the 24-hour solar day rhythm.[4] It would seem to be the same with plants, judging by a study of growth rhythms and sap flow in an orange tree: this varied through the course of the day, but when the tree was artificially maintained under constant light and constant temperature, the period shifted to a 25 hour rhythm, i.e that of the lunar day.[5] The tree preferred lunar time, in the absence of solar-day information.

This manner in which the cycle manifests in the plant realm could well improve scientific understanding of the matter. It makes sense to start investigating a phenomenon in its simplest mode of manifestation, not its most complex.

A rather important book here is 'Farewell to the Biological Clock' by G. Klein, which reviews the futile 20th-century biological quest for a 'chronon mechanism' or cellular clock that keeps the time for plants and small animals, chiefly of a 'circadian' or roughly-daily nature.[6] Klein argues that this is an illusion which now needs to be abandoned, because such a clock would have to have impossible properties such as being temperature-independent. Instead, he argues that cells have access to a substrate of lunar-day time, based on the 24.8 hour lunar day. They will tend to follow this, in the absence of night/day information. Normally the Sun coming out and then setting overrides such time-information and makes sure we all follow

---

[4] K. Endes and Wolfgang Schad, Moon Rhythms in Nature Floris Books, 2001 (English trans.)
[5] Millet and Moallem (2001) Growth rhyths and sap flow in Mandarin Orange tree, (in French) L'Arbre, 2000, Ed. M. Labrecque. 4th Int. Symposium on the Tree, Montreal, 97-103; cited in Zurcher and Holzknecht, 2006.
[6] Gunther Klein, *Farewell to the Biological clock*, Springer NY, 2007; with a Foreword by Peter Barlow, of Bristol university Biological Science department.

solar time. So, when you want to stay a bit longer in bed in the morning it might be your system yearning for that slightly longer lunar day!

The favored carrier of that time-information is Earth's fluctuating gravity field. The Sun's gravity pull is several hundred times stronger then that of the Moon here on Earth, but that of the Moon changes more every day, so it is the change in gravity field that living things are able to respond to and use for time-information.

To quote from a Foreword by British biologist Peter Barlow, underlying the night/day alternation:

> As a type of default time-keeping process, is the ability to perceive then respond to lunar forces. The default lunar-based response is, perhaps, often masked by more direct entraining stimuli, and only when these are absent can the default system make itself apparent.

We review below ten different aspects of plant growth: their metabolic rate, water absorption, growth rate, nutrient uptake, electrical activity, DNA production, fertilization, seed germination, wood hardness, and tree bud morphology, showing how in each case the synodic monthly cycle is pulsing through them. Of these, seed germination in relation to the monthly cycle has been the most thoroughly investigated, by at least half a dozen different investigators, as being a synodic lunar effect having a clear agricultural relevance. The researches of forestry scientist Dr Ernst Zürcher are here of paramount importance and may decisively affect future debate on this subject.

Metabolic Rate

In 1956, Professor Frank Brown began continuous monitoring of the oxygen metabolism of various plants in his laboratory. Their ambient physical conditions of light, temperature, pressure and atmospheric humidity were held constant. He chose root crops such as carrots and potatoes because they were accustomed to darkness, and so would be undisturbed by the uniform external conditions. After all, if plants adapted to a light/dark cycle were kept in continuous darkness, or continuous illumination, their cycles would soon become disoriented. With commendable patience, Brown clocked up over a million hours of potato time.[7]

Brown's potatoes, sealed from all light, were far from being in the dark about the motions of the luminaries in the sky. Potato metabolism followed

---

[7] F.A. Brown, 'The rhythmic nature of animals and plants', Cycles, April 1960 pp. 8192; 'Persistent rhythms of Oxygen Consumption in Potatoes, Carrots and the Seaweed Fucus' Plant Physiology, 1955, 30, p.280.

not only the daily cycle and the seasons, but also a distinct Moon-phase rhythm, peaking near full Moon and dipping near the New. For the fortnight centered on the Full Moon, he found that potato metabolism was 14% higher than that centered on New Moon, while for his carrots the figure was 11%. The diurnal cycle for comparison showed a mere 4-5% oscillation: i.e. the response to the Moon was considerably greater than to the Sun. His was not the only work to show this, as we shall see. During the cycle of the year, highest metabolism was in May, and least in October-November, the former being about twice the latter.

The lunar day lasts 24.8 hours, timing the rhythm of the tides, causing them to arrive nearly an hour later each day.[8] Brown reanalyzed his data for this daily cycle, discerning distinct lunar-day metabolism curves, not only for potatoes and carrots but also for algae, earthworms and salamanders.[9] Their distinctive feature was a sharp increase at moonrise. Why should potatoes be excited by the rising of the Moon? "Subtle, pervasive geophysical conditions." was the phrase used by Brown to describe the forces which he believed caused the changes shown in his experiments.

A fairly comparable experiment was conducted by Dr Eliane Graviou at Lyons University. She studied the small amount of oxygen which seeds used in their respiration when kept in darkness at constant temperature. With tomato and other seeds she found that maximum oxygen absorption tended to occur bimonthly, at Full and New Moon. Computer analysis showed that third and fourth subharmonics of 9.8 and 7.3 days were also present. In another series of experiments she measured root-lengths of three-day old seedlings, and found monthly rhythms present. Her results were published simultaneously in the International Journal of Biometeorology and the Journal of Interdisciplinary Cycles Research.[10]

Water Absorption & Tree Felling

Water uptake by plants ebbs and flows to the lunar cycle. After a three-year study using wheat and bean plants, Dr. Jane Panzer, a biologist at Tulane University, United States, reached this conclusion. She found that

---

8 A lunar day lasts 24.8 hours, given by: $1/24.8 = 1/24 - 1/(29.5 \times 24)$ the synodic month being 29.5 days.
9 For an update on the historic work by Frank Brown, showing how sea-shore creatures in his laboratory kept in tune with the Moon, including successful replication of his results, see: E Naylor, 'Marine animal Behavior in relation to lunar phase' (at Anglesea, U. of Wales), *Earth, Moon and Planets,* 2001, 85-86, 291-302.
10 E. Graviou, 'Analogies between Rhythms in Plant Material in Atmospheric Pressure and Solar-Lunar Periodicities', *International Journal of Biometeorology,* 1978, 22, p.103 (also *Jnl. of Interdisciplinary Cycle Res.* 1978, 9); see also T. Bryant, 'Gas Exchange in Dry Seeds,' *Science* 1972 178, p.634, for the diurnal cycle.

the water absorption peaked at full Moon, and that this was twice as much as in other quarters.[11]

Seed water absorption was studied in an experiment by Carol Chow and Frank Brown, of commendable simplicity. Performed in the open, unlike Brown's longer metabolism trials which had a closed environment, it involved taking a sample of dry pinto seeds from the same large sack of beans each day, shortly before noon. They were weighed, soaked in water for four hours and then re-weighed, and the percentage of water absorption ascertained. This was continued every day for two whole years.

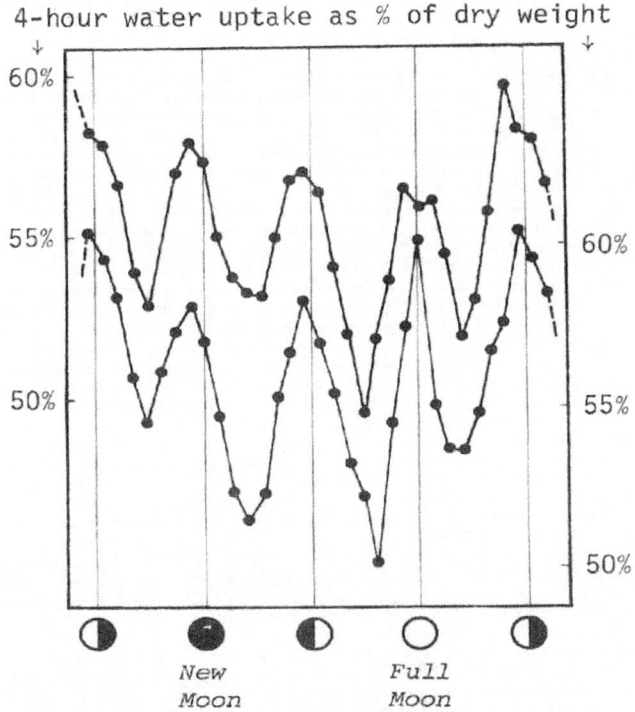

Fig 3.1 Brown and Chow (1973) Pinto bean % water absorption against lunar phase

The results showed a fine lunar fourth harmonic waveform (see figure 1), with absorption peaking at each quarter. The formation of a right angle between Sun, Earth and Moon caused the absorption of water by bean seeds to reach a maximum! Parallel experiments a thousand miles away at Woods Hole, Massachusetts using the same stock supply of beans showed a similar effect but with higher amplitude. Surprisingly, the same experiment

---

11 Dr Jane Panzer, 'Lunar Correlated Variations in Water Uptake and Germination in 3 Species of Seeds,' PhD Thesis Tulane University, Dissertation Abstracts International, 1976, 37,1077/8.

performed in a closed, temperature-controlled chamber gave a very different time-pattern (Figure 2). A single peak in absorption then appeared just before the Full Moon, showing no less than a 35% increase in water absorption at Full Moon compared to New! Thus, a natural rhythm was greatly modified by enclosure in an artificial environment.[12] Do trees follow any such cycle? Timber merchants in South America specify in their contracts the Moon phase in which trees are felled, on the grounds that timber felled at full Moon has too much moisture to be cut properly, while the drier wood felled at new Moon keeps better.[13]

The belief that wood cut at a new Moon is best goes back a long time.

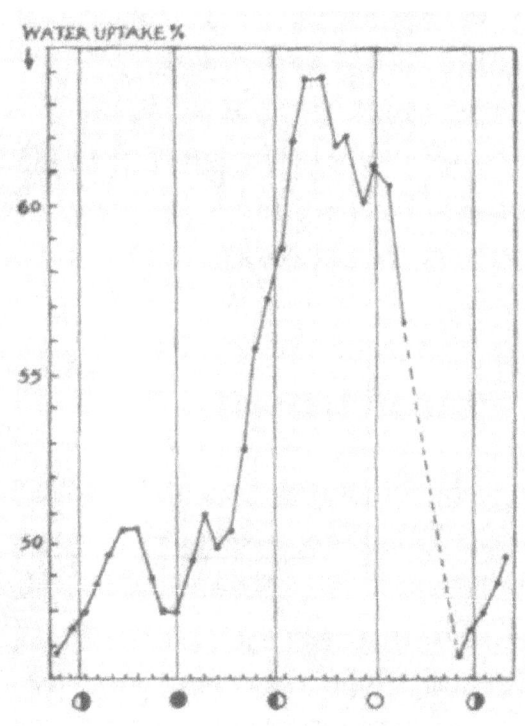

Fig 3.2 Brown and Chow (1973) Pinto bean % water absorption vs. lunar phase, performed indoors.

The Roman historian Plutarch wrote:

---

12 F. Brown & C. Chow, 'Lunar-correlated variations in water uptake by bean seeds', *Biological Bulletin*, 1973, October 145 pp. 265278.

13 E. Dewey, and O. Mandino, *Cycles-the Mysterious Forces that trigger events*, Manor Books, New York, 1973; E. and L. Kolisko, *Agriculture of Tomorrow*, 1978, p.12; J. Leon and N. Barrio, 'Influencia de las fases lunares en la atraccion de arboles trampas para escolitidos del genero Ips en pinares de Baracoa,' *Revista Forestal*, Baracoa, 1987, 17, 45-56. (Alluded to by Spiess, 2000, p.19 as verifying the traditional view that firs and pines felled before Full Moon suffered more from beetle attack than did those at New Moon.)

> The Moone showeth her power most evidently in those bodies which have neither sense nor lively breath; for carpenters reject the timber of trees fallen in the full Moone as being soft and tender, subject also to the wormes and putrefaction, and that quickly by means of excessive moisture.[14]

This traditional belief, that trees should not be felled while the Moon was Full, became enshrined in the French Napoleonic Code, for the forest conservators. This stimulated various derisive comments on the subject, but while researching for 'Planting by the Moon' back in the 1980s, its authors (Best and Kollerstrom) only came across one experimental investigation of the subject, in an Indian journal.[15] It was carried out, its author stated, to combat popular superstition on the subject. Water content of bamboo was measured every two weeks over six months.

The results, it was claimed, refuted the traditional peasant belief that sap rose with the waxing Moon. In fact the data showed a consistent and large amplitude 29.5 day cycle... with peak water content over New Moons!

Luna had the last laugh. In Mali the adage states 'Tu coupes l'arbre pendant la lune vide' (you fell the tree when the Moon is empty), when the wood is thought to be resistant against termites. We return to this question later, in a discussion of the 21st-century researches of Dr Ernst Zurcher.

Growth Rate

Growth rates of seven varieties of herbaceous plants were studied by botanist Giorgio Abrami. This was in the botanical gardens of Padua University, one of Europe's oldest centres of learning. These included Galanthus nivalis, the snowdrop, and anemone nemorosa, the wood anemone. Abrami was trying to relate growth to variations in daily temperature, but he kept finding days when the growth was not related to temperature as it should have been.

So, he went on to measure growth against the lunar month. He found that almost all of the plants studied grew according to a 29.5 day rhythm and also a bimonthly 14.7 day component, both related to the Moon's phases. The subharmonics of 9.7 and 7.3 day periodicities were also present, but these were weaker and not well linked to the astronomical cycle.

The blend of harmonics present varied from one species to another, i.e. the Moon was sounding a distinctive 'chord' in each species. Abrami found that 'conditions coinciding with the time of the Full Moon are less satisfactory

---

14 E.Tavenner, *The Roman Farmer and the Moon,* 1918.
15 C.Beeson and B. Bhatia, 'Effects of the Moon's Phase on Moisture in Bamboo,' *Indian Forest Records,* 1930, 2,12, p.241.

for maximum plant growth', a perplexing conclusion for followers of traditional lore.[16]

Abrami could have referred to an earlier and comparable study by Dr Arturo Lopez, of Madrid University in 1969. Lopez had found semi-monthly rhythms in the growth rate of several species of grasses and cereals, where the main periods present were of 14.8 and 7.4 days. Over several years, both with field and laboratory studies, he found these sine-wave effects, loosely linked to the monthly lunar cycle.[17]

The growth rate of trees is measured at a cellular level by mitosis, which is when a cell divides into two.

This was studied by some researchers in the Czech Republic, in vitro, that is with cells from embryonic trees. At the Mendel University Forestry department, they measured the 'mitotic index' which is the rate of cell division. Their tree cells would normally have 8% reproducing at any given time. Kept under constant conditions, in the dark and at a fixed temperature, they found that this rate of mitosis fluctuated according to the lunar month, peaking at Full Moon and reaching a minimum at the two lunar quarters. So, trees grow in tune with the Moon.

These investigators concluded that both the 29.5 monthly rhythm and a 14.8 day tidal rhythm was present in their data.[18] This result clearly demonstrates that there is no such thing as constant laboratory conditions, but instead there are tides, ever-changing, the Tides of Life.

Biochemistry

Nutrient absorption by plants in relation to this cycle was investigated by T.M. Lai, reported in the U.S. journal, 'Biodynamics'. While studying plant uptake of phosphorus from the soil, Lai found that it varied unaccountably. So, he set up an eight-month experiment using radioactively-labelled potassium and phosphorus, two vital nutrients in plant metabolism. Sudangrass seedlings in a growth chamber were allowed nine days for germination and growth, then assessed for nutrient absorption.

Surprisingly, Lai found a maximal phosphorus uptake around Full Moon with minimum around New Moon, whereas potassium uptake was the other way round. Two monthly cycles for nutrient absorption, for a metal

---

16 G. Abrami, 'Correlations between lunar phases and rhythnicities in plant and growth under field conditions', *Canadian Journal of Botany*, 1972, vol. 50.
17. D.Lopez, 'Ritmos de Periodo Largo en el Crecimiento de las Plantas' *Mem. Acad. Cien, Artes,* Barcelona, 1969, 396, pp169-218.
18 H. Vlasinova et al., 'The Mitotic Activity of Norway Spruce polyembyonic culture oscillates during the synodic lunar cycle,' Biologia plantarum 2003, 47, 475-6. For a graph of these results see Ch. 11.

(potassium) and for a non-metal (phosphorus), were 180° out of phase.[19] One of these is electropositive in solution, while the other is electronegative.

The biochemist Dr Harry Rounds at Wichita State University had been investigating stress hormones in the blood of mice and men, and was surprised to discern sharp peaks a day or so after the Full and New Moons each fortnight. These substances were cardio-acceleratory, and Rounds found a similar substance could be extracted from the leaves of various plant species, from Phaseolus to geraniums.

Sure enough, the cardio-acceleratory effect of these plant extracts changed sharply for a short period following each Full/New Moon - a conclusion not irrelevant to traditional advice that medicinal herbs should be picked at such times.[20]

Bio-Electric Fields

Each tree has its own electric field, and these ebb and flow to tidal rhythms. Professor Harold Burr, of Yale Medical College, discovered that when two electrodes were put into the trunk of a tree, one above the other, a small current measured in millivolts flowed one way or the other. Burr was surprised to find that fluctuations in this current were the same for all the trees tested over a large area. A maple tree was selected and its trunk potential monitored over fifteen years.

Surprisingly, he found that the voltage was not systematically linked with such obvious environmental factors as temperature, humidity, atmospheric pressure, rainfall or daylight. It did, however, vary with air and earth potentials, and responded to the course of the day, peaking in the afternoon, and to the sunspot cycle and turmoil on the Sun's surface. During the year's course, positive and negative voltage peaks occurred at the equinoxes.

A student of Burr's, R. Markson, analyzed the years of data in some detail, comparing its 27.3-day cycle, which was solar, as being the mean sunspot rotation period, with the 29.5 day lunar cycle. His analysis showed that, in general, the lunar periodicity was stronger in the data than solar periodicity. His computer also picked out overtones of the fortnightly period, i.e. rhythms whose periods were whole-number divisions of 14.7 days.[21]

---

19 T.M. Lai, 'Phosphorous and potassium uptake by plants relating to Moon phases,' Biodynamics (US Publication) summer 1976, pp.1-15.
20 H. Rounds, 'A Semi-lunar periodicity of Neurotransmitter-like substances from plants,' *Physiologica Plantarium*, 1982, 54, 495-9. (More recently, some very general comments of doubtful value are found in Ian Cole and M. Balick, 'Lunar Influence: Understanding Chemical Variation and Seasonal Impacts on Botanicals,' HerbalGram, The Journal of the American Botanical Council, (online) 2010; 85, 50-56.
21 H. S. Burr, 'Diurnal potentials in the maple tree,' *Yale J. Bio. Med.* 1945 vol. 17, p. 727. R. Markson, 'Geophysical Influences in Biological Cycles', *Journal of Interdisciplinary Cycles Research*, 1972, 3,134; a more extensive account of Markson's work is found in an

A PhD study has replicated the Burr finding, focusing on the quiet winter months as being the best time to register the lunar-phase electrical waveforms: by Kurt Holzknecht at Innsbruck, his thesis on 'Electrical potential in the sapwood of Norway spruce' showed in 2001 how the potential along a tree trunk varied with lunar phase and the 'gravimetric tides' (which respond primarily to the moon's diurnal variation).[22] Zurcher has co-authored a paper with Holzknecht on this topic, subjecting the results to Fourier analysis.[23]

DNA

The 'golden helix' of life, the self-reproducing molecule, contains a lunar pulse in plant material. Biochemists from the University of Paris reported this result at the Amsterdam Congress of Geo-Cosmic Relations of 1989. Their graphs were a model of inscrutability, so I cannot give much detail, but they discerned two types of plant nuclear DNA mass, which they called type a and type b, and X-ray crystallography showed that the type 'a' tended to cluster around type 'b'.

The type 'a' grew more during New Moons, while the type 'b' appeared in larger quantity during Full Moons: 'it seems that the plant program has the possibility to choose between the two forms, a and b, depending on the environmental conditions...[For the 'a' form] expression of the program...is oriented towards storage of carbohydrates...[for the 'b' molecule] the expression of the program is oriented towards plant growth and flowering.'

Having two types of DNA linked with each end of the lunar cycle could, if confirmed, establish the whole subject on a firm scientific basis.[24]

Fertilization

Bees are well adapted to the lunar cycle in their flight activity, as revealed in studies by M.G. Oehmke, a biologist at the Goethe University of Frankfurt. While researching the time-sense of honey bees, Oehmke started to notice a pronounced monthly periodicity in his data, though the exact form of this adaptation varies from one species to another. He fixed

---

appendix to *Blueprint for immortality* by H.S. Burr, 1978, 'Tree potentials and external factors.' For how Professor Burr was able to publish this material while holding his chair of Anatomy at Yale, see Lyall Watson, *Supernature II* 1986, p.113-4; for a summary see Lyall Watson, *Supernature* 1973 pp.81-88.

22 . K. Holzknecht, 'Electric potential in the sapwood of Norway spruce... and their relationship with climate and lunar phase,' PhD Innsbruck U. Botany Dept., 2002; cited in Zurcher, 2006.

23 . E. Zurcher and K. Holznecht,, 'Tree Stems and Tides,' Schweitz z. Forstwez, 2006, 157, 185-190.

24 M. Rossignol et al., 'Lunar Cycle and Nuclear DNA variations in potato callus' in Geo-Cosmic Relations (ref.2) pp.116-126.

electronic counting devices onto his beehive entrances and thereby monitored whole colonies of bees, counting thousands every day, through the course of a year.

One species had a semi-lunar flight activity over the summer months, peaking at quadratures (i.e. quarter Moon positions), while another species (Carnaca bees) had an activity cycle of double this length, a distinct monthly rhythm peaking at the New Moons. The bees were then at least twice as busy as over the Full Moons! The character of this rhythm would alter at the equinoxes.

Oehmke took his 'Carnaca' bees indoors over the winter months and gave them a continuous light regime, which caused them to undergo a 180° phase reversal, so that they were now coming out most often at the Full Moon.

This result is reminiscent of what Brown and Chow found with his bean seed water absorption, where the lunar cycle altered upon taking the experiment into his laboratory. Oehmke explained these phenomena by averring that his bees were detecting monthly fluctuations in the Earth's gravity field.[25]

Confirmation of these results has found a fortnightly or tidal rhythm in the glycemia of the honey bees, peaking at both the Full and New Moon positions.[26]

This research was done in Morocco on a government grant, and follow-up studies are scheduled.

Germination

The opinion received is that seeds will grow soonest if they be set in the increase of the moon. - Sir Francis Bacon.[27]

Sir Francis Bacon proposed an experiment to test this durable piece of folklore, whereby seeds would be sown in pots at different lunar phases. Batches of the same seed stock should be sown in similar soils and shielded from the weather, he wrote, "lest the difference of the weather confound the experiment."

Three centuries rolled by and finally someone performed the experiment: L. Kolisko, who first published her results in 1929, then again in 1935 and

---

25 M.G. Oehmke, 'Lunar Periodicity in Flight Activity of Honey Bees,' *Journal of Interdisciplinary Cycle Research* 1973, 4, 319-335.
26 Mohssine et al., 'Lunar Phase Influence on the Glycemia of Worker Honeybees,' Chronobiologia 1990, 17, 201-7. For an earlier US study of bee pollination in relation to lunar phase see W.B. Kerfoot, 'The lunar periodicity of Sphecodogastra texana, a Nocturnal Bee' Animal Behavior 15, 1967, 479-486.
27 Francis Bacon, Sylva Sylvarum (1627) in J. Spedding & R. Ellis, Eds, 'The Works of Francis Bacon', Vol.2, 1887, p.636.

1942 .[28],[29],[30]) She germinated wheat seeds in beakers of soil and grew them for two weeks, watering every other day. Eight batches at a time were set at each lunar quarter throughout the year, measuring the lengths of the first and second leaves.

Her experiments were carried out in a greenhouse, without temperature regulation, from 1926-34. Her published graphs are not easy to interpret, but seem to indicate that maximal growth occurred in the waxing as opposed to the waning half of the lunar month, except during some winter months when this was reversed.

Her general conclusion was that the days prior to the Full Moon were optimal for seed germination. Kolisko's 'Agriculture of Tomorrow' was first published in 1940, which stated as a summary of her experimental findings:

> The maximum growth is always reached during the waxing moon-period, from New Moon to full moon. That seems to be a law.

Field experiments by Mather at the John Innes Horticultural foundation in 1940 showed a consistent fifteen percent yield increase for maize and tomatoes sown in the second lunar quarter (see Appendix I), confirming the very hypothesis that was being test - that which Kolisko had just published in her Agriculture of Tomorrow.

Despite this confirmation, the author of the report was quite dismissive of his results,[31],[32] and subsequently a review in Nature echoed his skeptical tone.[33]

---

28 L. Kolisko, 'The Moon and the Growth of Plants,' 1936, 1938,1978. A shorter account 'Der Mond und das Pflanzenwachstum' was published in Stuttgart in 1933: her experiments were 'first published in 1929, which stimulated experiments by farmers.' They extended over an eight-year period, from 1927-34. They were never adequately reported, in that the graphs were hard to interpret, as their axes tended to be unlabelled.

29 G. Husemann, 'Lili Kolisko - Her Life and Work,' Archetype, Science Group of the Anthroposophical Society in Great Britain, Editor David Heaf, September 2001 pp.31-48 (trans. from the German by Heaf).

30 E. and L. Kolisko, 'Agriculture of Tomorrow,' 1939, 1978, Ch.2 (one can only regret her altering her name, from Elizabeth to Lily).

31 K. Mather & J. Newall, 'Seed Germination and the Moon,' *Jnl. Roy. Hort. Soc.* 1941, 66, 358-66. Results were cited as, days after sowing when 50% of the seeds had germinated, to the nearest day, e.g. 5 days, hardly a sensitive or reliable measure (See Appendix I).

32 M. Mather, 'The Effect of Temperature and the Moon on seedling growth', *Jnl. Roy. Hort. Soc.* 1942, 67, 264-70. Hans Eysenck and David Nias erroneously remarked, on Kolisko's plant-growth experiments: 'Amazingly, there has been no systematic follow-up to this work' ('Astrology: Science or Superstition?' 1982, p.167). Mather's experiments were quite systematic, however deluded his conclusions may have been (Appendix 1).

33 C. Beeson, 'The Moon and Plant Growth,' *Nature* 1946, 158, pp.572-3.

Through April to August of 1940, Mather sowed twenty-four rows per month, for three different trials: Tomato, Maize I and Maize II. Two rows of each kind were sown in each trial, ie six at a time, and were sown two days before each lunar quarter.

His data have been reanalyzed by the present writer, first standardizing the sets by adjusting the mean of each to 100. Combining the total of 78 rows sown by Mather in the four lunar quarters gave:

Full Moon sowings (n=18)   112.7±23 % of mean

Others (n=60)   97.3±27 % of mean

Where 'n' is the number of rows: a precise confirmation of the effect found by Kolisko, and published by her in the very year in which Mather was conducting his experiments.[34] On average, the Mather field trials indicate an average second-quarter excess, i.e. in the week leading up to the Full Moon. after a month's growth, of 15% for crops of tomatoes and maize.

Rightly interpreted, the Mather field trials offered a significant answer to the question posed three and a half centuries ago by Francis Bacon. Issues are here raised concerning data analysis, and the separation of seasonal (solar) and monthly (lunar) factors. I offered this re-analysis to 'The Garden', the present-day version of the journal in which Mather published, but it was declined (see Appendix 2).

A lunar cycle in the germination rate of seeds was described in 1967 by M.G. Maw.[35] Initially, Maw was investigating the effect of positively and negatively charged air ions on germination rate; however, he found that in some cases this seemed to increase growth rate whereas in other cases it did not, and from this he was led to observe that germination was influenced by the lunar cycle.

In experiments performed between 1963 and 1966, he grew batches of cress seed in distilled water, estimating the growth rate after a 5-day period (using the mean stem length per batch). He found that the control group, which received no ionized air, usually grew most over the time of the Full Moon and least over the New Moon. However the opposite effect tended to occur with seeds exposed to negative ions. As the Moon has been shown to affect the flux of ionized particles entering the earth's atmosphere differentially in the course of its monthly cycle, Maw suggested that some ionic mechanism could here be involved.

---

34 The reanalysis (Appendix 1) was submitted in 1996 by the present writer to *The Garden*, the present-day name of the Journal of the Royal Horticultural Society, but not accepted.

35 M.G. Maw, 'Periodicities in the influence of air ions on the growth of garden cress,' *Canadian Journal of Plant Science*, 1967, 47, 499-505.

A six-month trial of wheat germination by the present writer had features in common with the Kolisko and Maw procedures.[36] Half a dozen batches of wheat seeds, twenty or so at a time, were germinated on floats in beakers so that the seeds were kept sufficiently moist to germinate. They were grown in the dark at uniform temperature for a week, being set on the same day each week. This procedure requires no awareness of lunar phase on the part of the person performing the experiment.

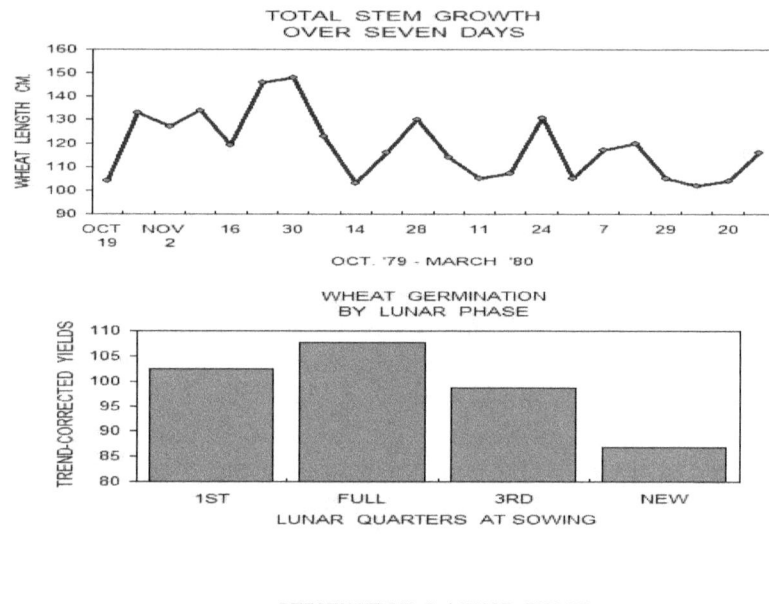

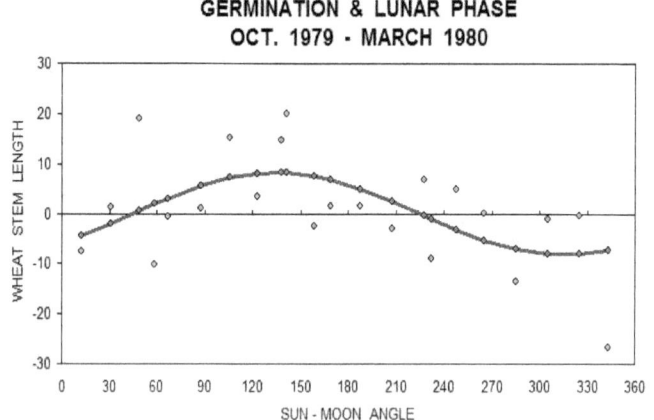

---

36 N. K., 'Wheat Germination and lunar phase, a pilot study,' Correlation, May 1984 25-31.

Fig 3.3 Mean wheat stem length after 7 days' growth, over 6 months 1979-'80; 3.4 divided by lunar phase; 3.5 plotted by Sun-Moon angle, with best-fit sinewave.

Total stem length per batch was measured at the end of each weekly trial, ie the sum of the lengths of all the wheat seedlings that had grown per batch. The effect was roughly equal on germination and stem growth rate per batch, so it was convenient to combine these two into the one measurement. Thereby each week one ended up with half a dozen total stem-length measurements. These displayed a monthly waveform, in which Full Moon germinations did some 14% better that New Moon ones, ie the waveform had an amplitude of 7%. As Kolisko had found, the peak of the waveform was shortly before the Full position

There was a suggestion that lunar latitude was relevant, whereby times near to the node were less favorable to growth. The (Ewell) college would not accept the proposal as suitable for further investigation and the project had to be terminated. This is the experiment I would most like to replicate, as it has a primal simplicity. Other studies have indicated that levels of solar activity influence seed germination[37,38,39], which should also be taken into account. The more recent germination trials by Zürcher using tree seedlings are described below.

Tree Buds

Fortnightly rhythms in tree bud morphology have been recorded in great detail by Lawrence Edwards, as described in his book, The Vortex of Life.[40] As a school mathematics teacher, he had for years been investigating the kind of geometry that could model egg and bud shapes, and then in 1982 he realized that the pattern of this form was pulsing slowly in time, to a tidal rhythm. To quote from a review of his book in the New Scientist:

> Edwards shows that dormant buds are not as quiescent as they appear. Measuring them daily, he found that buds pass through a roughly fortnightly cycle of shape changes. During the cycle, buds relax their form slightly, as if beginning to open, and then

---

37 B. Cummins, 'Correlations between periodicities in Germination of Chenopodium Botrys and Variations on solar Radio flux,' *Canadian Journal of Botany*, 1967, 45, 1105-1113. Cummins, at Ontario University Botany Dept., found 'highly significant correlations between fluctuations in germination and solar radio flux (10.7 cm wavelength, indicative of sunspot activity).'
38 B. Cummins, ref. (37).
39 G.Abrami and G.Piccardi, 'Seed Germination as a Biological test for the study of Fluctuating Phenomena,' Jnl. Int. Cycle Res. 1973,4,267-82. This concluded that the Geomagnetic field variations had influenced germination.
40 L. Edwards, The Vortex of Life: Nature's Patterns in Space and Time, 1993.

close up again. The pattern is widespread, appearing anywhere from the leaf buds on beech trees to the flower buds on primroses, and the cycle is roughly synchronized between buds on the same plant.[41]

Edwards has also investigated the manner in which high-voltage cables nearby seem to disrupt this periodicity. In his view the rhythm is not quite tidal but slightly shorter. A later section returns to this topic: over the four months or so of the winter, when the buds on trees are formed and apparently doing nothing but waiting for spring to arrive, to what rhythm are they oscillating?

Edwards' work reminds one of the more recent work of Fritz[42] who showed that the formation of new leaves was strongly linked to the synodic lunar rhythm, with a positive Full-Moon effect, but with 'offbeat' periods giving formation maxima at the New Moon.

From Kolisko to Zürcher

The findings of Ernst Zürcher started to appear towards the end of the 20th century.[43,44,45,46,47,48,49] From the viewpoint of biological science, no other 20th-century experiments demonstrate so clearly the Moon's influence upon the plant realm, and in a manner of clearly practical significance. They involve well-performed, large-scale experiments by professional foresters. There are some four stages of his work: the trials of tree seedling germination in Africa (1989-91), then the tree-trunk water content trials in Switzerland, then his startling findings on tree-width oscillations as

---

41 New Scientist, 13 November 1993 p. 17, reviewed by Stephen Day.
42 J. Fritz, 'Studies on the influence of the synodic moon rhythm on the growth of radish', 1994 Kassel U. Germany (source: Zurcher).
43 E. Zürcher, 1992 'Rhythmicities in the Germination and Initial Growth of a Tropical Forest Tree Species' (in French), *Journal Forestier Suisse*, 1992, 143, 951-966.
44 E. Zürcher, 'L'Arbre, Biologie et Développement', *Troisième Colloque International Monpellier* 1995, Ed. Edelin, C., 150-164.
45 E. Zürcher et al., 'Tree Stem diameters fluctuate with tide', *Nature* 1998, 392, 665-6 (the fluctuations were several hundredths of a millimetre).
46 E. Zürcher, 'Lunar-Related traditions in Forestry and Phenomena in Tree Biology' (in German), *Journal Forestier Suisse*, 2000, 151 417-424.
47 E. Zürcher and Daniel Mandallaz, 'Lunar Synodic Rhythm and Wood Properties,' *L'Arbre* 2000, 4th International Symposium on the Tree, Montreal, 2001, 244-250; Zurcher et.al., 'Looking for differences in wood properties as a function of the felling date: lunar phase-correlated variations in the drying behaviour of Norway Spruce and Sweet Chestnut,' *Trees* (2010), 24:31–41.
48 E. Zürcher, 'Lunar Rhythms in forestry Traditions - Lunar-Corelated Phenomena in tree biology and wood properties,' *Earth, Moon and Planets, an International Journal of Solar System Science*, Kluwer Academic, Holland 2001, 85, 463-478.
49 K. Holzknecht and E Zurcher, 'Tree stems and tides – A new approach and elements of reflexion' *Schweiz. Z. Forstwes.* 2006, 157, 185-190.

published in the science journal Nature (1989), this being the first significant article it has published on the subject, and finally a look at electrical oscillations in tree-trunk voltages, as Burr had done earlier.

"These fundamental things must be simple" was a saying of Lord Rutherford, the 20th-century British nuclear physicist. The Zürcher experiments differ from most published science in that they have sought to investigate the phenomenon at the simplest level, and if they become memorable it will be, I venture to suggest, because of this. Half a century later, Kolisko's basic concepts have finally been validated, in Zurich, not far from where she worked in Stuttgart, by Ernst Zürcher, who works in the department of Wood Sciences at the Swiss Federal Institute of Technology.

While working for four years as project-leader of the Swiss Development Cooperation in Rwanda, Zurcher realized that this was an ideal place to investigate tree chronobiology because of the almost constant day-lengths and temperatures throughout the year. The Institute des Sciences Agronomiques du Rwanda was aided by a Swiss Inter-cooperation department in Berne for a project, where twelve sets of sowings were made in 1990-91, of four different tree species, sowing prior to Full and New Moons. How many days the seeds took to germinate were counted, plus the mean height which the trees reached after four months of growth.

The Full-Moon half of the data scored consistently better for both tree germination and initial height (see Appendix 1). Zürcher thus confirmed Kolisko's findings of better emergence and subsequent growth in the Full-Moon sown half of the data, for trees in the tropics. (He surmised that the growth hormone cytokinine was more strongly present in the Full Moon sowings and cited an earlier German study on algae that had shown this).

The maximal height of the trees after four months of growth showed 'the Full Moon sowings always coming ahead of the new moon ones.' In this area, folklore becomes science when we become able to describe the process by mathematical rhythms present in the data. Four years after his Rwandan experiment, an independent work on tree seed germination and initial growth of four African species was carried out, in Mali, using the same method, and showed the same result.[50]

As Zürcher rightly concluded;

> These trials make clear, for the first time in trees or shrubs, the existence of a real phenomenon, often mentioned in traditions or issuing from empirical experience, consisting of a link between lunar phases and the behavior at germination and during initial growth.

---

50 N. Bagnoud, 'Rhythmicities in the Germination and the Initial Growth of 4 Tree Species of the Soudano-Sahelian Zone. Moon phase trial', Berne: cited in Zurcher, 2001.

Kolisko in her 'Agriculture for Tomorrow' published in 1939 had written, 'That seems to be a law,' alluding to her finding that seeds germinated better on days prior to the Full Moon. Her work was generally met with polite skepticism and she ended up as a bit of a hermit. Science is a public enterprise and it depends upon independent replication of experiments, as has now been done. Having the trials in Africa was no doubt helpful, as things grow quickly there. Zürcher viewed his findings as significant in terms of the need to establish new forestry plantations, as the Kyoto Climate Conference had called for in 1997.

Over the winter of 1998-99 Zürcher and his colleague chose six felling dates near to syzygy (the Full and New positions), on each of which they felled five large trees of Norway spruce, from the experimental forest of the ETH at Zurich. A series of wood samples were then taken from each tree and dried. This is not an easy experiment to perform under standardized conditions. It was performed under the aegis of the Department of Forest Sciences of the Swiss Federal Institute of Technology. That such an institute was prepared to condone such an experiment shows how greatly the climate of opinion has shifted. Has there been any other case where folk traditions enduring for two millennia - since Pliny in ancient Rome, at least - were decisively confirmed by a scientific experiment.

He tested the moisture-content of the spruce-tree wood-samples over three lunar months at fortnightly intervals, using a procedure that involved careful oven-drying of the wood.

The overall final mean values of density were: 0.46 grams per cubic centimeter for fellings made near the Full-Moon and 0.51 for those towards the New Moon - that's a huge difference of around fifteen percent, with New Moon wood being denser, dryer and more compact.

A debate that had rumbled on through centuries was settled by the experiment. From millennia of folk-tradition there emerged a scientific fact: for wood used for construction, the optimal tree-felling time is before a Christmas New Moon (i.e. December-January).

They also investigated the compression strength of sapwood from felled trees and found a similar result. The New Moon-average of 47 N/mm2 (this is a measure of pressure, newtons per square millimeter) was found to be substantially greater than the Full Moon-average of 42 N/mm2. As Zürcher commented:

> These felling-date-related variations are astonishing at first sight for our current knowledge of wood physics and ask for a formulation of further working hypotheses.

The relative density of tree sapwood around the edge of the trunk varied more with the lunar pulse than did that of the heartwood in the middle of the trunk.

Lunar Phase Rhythms in Plant Growth

Zürcher then turned his attention to the rhythmic diurnal fluctuation of tree-trunk diameters, the idea being that these would reflect any diurnal patterns of water absorption. Through giving a report of his earlier work at a Montpelier symposium, he came across some Italian researchers, who were detecting a 25-hour rhythm in tree-width thickness, and were publishing it without any hypothesis linked to lunar cycles.

He collaborated with a French colleague in reanalyzing this data. The trees had been kept in darkness and under constant temperature during the experiment, which made the lunar effect more evident: the tree-trunks were expanding and contracting with the diurnal position of the Moon! Minute variations in tree stem diameters were co-varying with the Earth's gravitational field.

This field pulses on a daily basis, as the Moon passes overhead, although it is quite a complex business. The minute gravity-field fluctuations are measured in 'gals' after Galileo who first struggled with the question of the tides and Earth's motion. An oceanography department supplied the gravity-data. To quote from Zürcher's 1998 article in *Nature*:

> The diameter of tree stems growing under open and controlled conditions undergoes rhythmic fluctuations independently of daily periodic factors such as light, temperature and humidity. We find a strong correlation between these fluctuations and the timing and strength of tides. This correlation suggests that the Moon is influencing the flow of water between different parts of trees.

This cycle 'usually appears as a double-peaked wave with a period of about 25 hours.' It was still found in cut-off trunk sections 'as long as the cambium is alive.' Even a cut-off bit of trunk shows this pulse! It seems to me that science is about matter, about the material universe, and this experiment of the Italian researchers is about matter, *mater, materia*. This seems to be a bio-gravitational effect.[51]

In the course of making this historic announcement in Nature, Zürcher remarked in passing that a book *Biologie des Mondes, Mondperiodik und Lebensrhythmen*[52] had reviewed lunar influence in about six hundred animal and plant species. The Germans have a reputation for being thorough in these matters! Zürcher's research seems to have been inspired somewhat by the prevalence of lunar-based traditions amongst foresters, and how consistent they were:

---

51 But, for a failure to replicate the effect Zurcher described in the *Nature* article (ref. 132), see T. Vesala et al., 'Do tree stems shrink and swell with tides?', *Tree Physiology*, 2000, 20, 633-635.
52 K. Endes and Wolfgang Schad,*'Moon Rhythms in Nature* Floris Books, 2001 (English trans.)

> The general rules governing the felling of trees are in accordance right across the continents: whether in the alpine arc (Hauser, 1973), in the Near East, in Africa, India, Ceylon and Brazil, or in Guyana, all these traditions seem to be based on matching observations (Broendegaard, 1985; other sources, see Zürcher 2000). It should be noted that in the past, people had more time and more peace and quiet to observe: it must even have been of vital importance to them.[53]

His work alludes to various significant German studies, with which the present writer is not altogether familiar. In the alpine forest and also in Pays d'En Haut, some wood wholesalers and musical instrument-makers would ask to be present on the felling-date in order to guarantee the wood quality. The wood had to dry especially well and this helped its acoustics. Lunar phase and also some zodiac lore were considered important. Full-Moon felled wood was said to be the best for firewood, the wood then having a lighter consistency.[54]

Zurcher also cited a successful, internationally-known family timber enterprise near Salzburg, where slow-grown mountain forest trees are felled at new Moon and additionally during a 'dry' sign such as Sagittarius. This firm is able to guarantee high-quality building timber.

If forest tree trunks are varying substantially in their cellular binding of water, to a monthly rhythm, then this is going to have huge implications, for our very notions of life itself, and goes beyond merely aiding lumberjacks to get good timber. Why should living matter have this property, and how could it possibly work? This is a tidal rhythm, but of double the tidal period, as tides ebb and flow twice-daily. In the heart of a tree, in the heart of a forest, a rhythm pulsates, a rhythm that is astronomical. That is not mysticism, it's biology.

Trees are displaying a very total response to the interweaving rhythms of Sun, Earth and Moon. Through their water processes, their trunks are responding to the matrix of these rhythms. This relates to earlier findings cited in this chapter about tree electricity, and it also seems connected with Lawrence Edwards' work on the shape of buds, and how they are responding to a fortnightly rhythm. Edwards had to grapple with the question as to whether he could really believe that all buds during the winter months were oscillating in silence to this fortnightly rhythm; Zürcher's work may help us towards answering this.

---

53 Zürcher and Mandallaz, ref. (47), p.464.
54 For traditional lore on timber-felling, see *Moon Time, the art of harmony with nature and lunar cycles* by J. Paungger and T. Poppe, 1995, Ch.4 (trans. from the German).

However, and in contrast with Edwards' research, Zürcher's is of immediate practical utility. His findings have been presented at international forestry symposia.

We have been surveying evidence for a kind of music within the plant realm. The effects are of large-amplitude, and indicate a silent symphony ongoing, differentiated from one species to another, changing its tune through the month: the tides of life.[55]

A plant grows through various rhythms, unfolds through a time structure, in tune with the cosmic process. A dry and lifeless orb modulates water- and life- processes in a plant: that is the mystery.

These scientific, university-based experiments provide a background to our subject, one which makes the utilizing of a lunar gardening calendar seem mere common sense.

After all, it would be bizarre if all these rhythmic processes were proceeding in their highly energetic manner each month, without the application of a lunar planting manual being a straightforward matter, for utilizing these energies.

One cannot readily infer as to what principles should be contained in such a manual from the foregoing, yet the data strongly implies its potential usefulness.

---

55 NK., 'Plant Response to the Synodic Lunar cycle, a Review,' Cycles (Pittsburgh, US) 1980, 31, 61-63.

4. Four Elements in the Zodiac

It is evident, however, that all bodily things derive their origin from the earth, and similarly also their being, according to the laws of time, through the influence of the stars or planets - sun, moon and the rest - together with the four qualities of the elements which agitate them ceaselessly. By this means each and every growing and fruitful thing is brought forth with the type and form proper to its own substance. - Splendor Solis, 1598[1]

A plant begins its life, its existence in time, when moisture expands the seed and its DNA coils to begin to duplicate. That is the moment of beginning, when the firm structure of the seed breaks down into a chaos. Before that, it was in a more or less static condition. At that one moment, does the quality of Time then affect it, determining to some degree such things as its future form, fertility and edibility? If so, this would resemble the astrologer's claim about the unique moment of birth, that it has a significance for the future life.

To Claudius Ptolemy, composing an astrological treatise in the second century AD, these things appeared fairly self- evident:

... the germination and fruition of the seed must be moulded and conformed to the quality proper to the heavens at the time.

This was something which, he could affirm matter-of-factly, 'all would judge to follow,' adding moreover that 'the more important consequences signified by the more obvious configurations of sun, moon, and stars are usually known beforehand, even by those who inquire, not by scientific means, but only by observation'.[2] On this basis, it was mere common sense for the farmer to prognosticate by assessing the condition of the heavens.

Ptolemy perceived an analogy between the moment when a seed began to germinate and the moment of human conception. He explained why these moments had their special significance, because of this view about the moment of germination:

For to the seed is given once and for all at the beginning such and such qualities by the endowment of the ambient; and even though it may change as the body subsequently grows, since by natural

---

[1] Salomon Trismosin, *Splendor Solis,* Germany 1598, trans. Joscelyn Godwin, Phanes Press 1991, 19.
[2] Claudius Ptolemy, Tetrabiblos Loeb Classical Library 1980, I.2.9. Ptolemy was an Egyptian in Alexandria, and wrote treatises on geography, astronomy and astrology.

> process it mingles with itself in the process of growth only matter which is akin to itself, thus it resembles even more closely the type of its initial quality.[3]

As one commentator on Ptolemy has observed;

> In this way the heavenly configuration has a powerful effect at the seed's uniquely impressionable moment, the instant of fertilization.[4]

Nowadays however we no longer find these things nearly so self-evident, as did Ptolemy in the second century.

Should we believe that there are certain periods in the month optimal for sowing a crop, which are likely to affect the harvest obtained from it? If so, can we hope to ascertain how such determinative effects might work? Such a claim might take the form, that for trees a good aspect to Saturn is beneficial, or that for annual crops the lunar nodes should be avoided. This chapter focusses on a view that developed in the 1950s, based on the traditional four elements, cast as a zodiac-pattern in the sky. It will give a historical perspective and review published evidence for and against it. It presents a theory, a theory of celestial influence.

This turns out to be a rather different issue from the Moon-phase rhythms we have surveyed hitherto. The lunar phases each month encourage growth, so that any plant receives that monthly stimulus regardless of when it was planted. Lunar gardening calendars often advise that root crops should be sown near the new Moon, and other crops near the full Moon, but evidence supporting this is scant: trials usually indicate that the Moon's phase at the time of sowing is of little relevance for achieving sizable yields.

A theory originated by a German woman of peasant stock, Frau Maria Thun (pronounced 'Toon'), is the basis of the sowing calendar used by many Biodynamic farms. She envisaged the constellation in which the Moon is standing at the time of sowing as affecting subsequent growth, depending on what was traditionally its "element". Thereby a rhythm was generated in plant growth by the Moon's motion in relation to the fixed stars called a "sidereal" or star rhythm.

There is a long tradition of assigning the Zodiac signs to the four elements. For example, Taurus, Virgo and Capricorn are the three signs (of corresponding constellations Bull, Virgin and Sea-goat) associated with the element Earth.

Chapter 7 will discuss the difference between signs and constellations. Suffice to say here that when the four-elements were being placed into the

---

3 Op cit. ref (2), III.1.225.
4 Geoffrey Cornelius, *The Moment of Astrology,* 1994, p.91. This contains an extensive discussion of Ptolemy's 'seed' argument.

zodiac in the early centuries AD, the two schemes were to a large extent coincident. We use the convention of using latin names for signs and English names for the constellations.

According to the Thun theory, root crops such as carrots or potatoes will tend to grow best when sown when the Moon is passing in front of one of these three Earth-element constellations.

Biodynamic farmers call these "root-days". The Moon enters one of these three constellations every nine days, because they are at 120° intervals in the circle of the Zodiac. Biodynamic farmers call "leaf days" those days when the Moon is passing in front of the water constellations, Scorpio, Cancer and Pisces. Leaf crops, such as lettuce, are sown on these days. Flower and fruit seed crops are also sown when the Moon is in front of air and fire constellations respectively.

So the traditional four elements are assigned to four types of crops: root, leaf, flower and fruit seed. This is summarized in the Table.

There is an attractive simplicity to this model, in that it does not involve differentiating the separate constellations beyond this elemental pattern. In this respect it differs from more traditional 'astrological' guides, which generally affirm that Libra for example is good for sowing flowers (as being ruled by Venus) or that Taurus is especially fertile.[5]

The model leaves the separate signs undifferentiated, not viewing any particular zone of the zodiac as especially fertile. This should appeal to antipodean farmers since, while Aries and Taurus have traditionally been linked to Spring's burgeoning growth, this cannot be experienced Down Under.

The star-zodiac has no inherent connection with the seasons of the year. The sole distinct constellation-attribute used by Biodynamic farmers is the (rather curious) linkage of the Lion with seed quality:[6] they sow and/or harvest in this constellation when seeds are to be gathered.

Thun has been publishing the results of her investigations into "sidereal" Moon-rhythms since 1962. Her method involves sowing twelve successive rows of one type of crop during a sidereal month, sowing each time the Moon passes the middle of a constellation.

---

5 Louise Riotte, *Planetary Planting* 1990, Astro-computing Services US; earlier, this view was expressed by Dr C. Timmins, *Planting by The Moon*, 1939 Aries Press Chicago (I haven't seen a copy); see Dean 1977 p.64 for a summary of Timmins' views.

6 Thun, *Work on the land and the constellations* 1991; *Biodynamics, New directions for Farming and Gardening in New Zealand*, 1989, p.136. As another instance of Thun using a specific constellation, she advised: 'Control slugs and snails, if necessary, when the Moon is in front of the Crab' (Working with the Stars, 1992, p.43). The 'watery' constellation of the crab is here envisaged as having some attunement to these moist pests.

Harvesting is usually done to a similar rota one row at a time, so that each row has the same growing period. The weight of the yield of each row is recorded, and the results appear through this measure.

Symbolism

At the heart of the Biodynamic calendar is a fourfold structure. Let us arrange the four elements, not in their zodiac sequence, but in an order of decreasing density: starting with the solid earth, then into water and air, ending up with fire. Traditionally the first two of these elements had weight, i.e. tended downwards, while the other two had rather the opposite and tended to move upwards, thereby maintaining a balance between gravity and levity. Thun envisaged these four elements within the fruits of the earth:

Fig 4.1 The Thun four-element theory, linking four crop types & zodiac element.

Earth: the seed germinates in the dark, stretching its roots through the soil.

Water: the stem and leaves grow upwards, as water flows up through the stem and out through the leaves.

Air: the flower opens to the sky, breathing out its fragrance and attracting bees and butterflies.

Fire: the heat of the summer matures and dries the plant, after its flowers have gone, concentrating its being into the potency of seed.

One's first reaction is: can anything be that simple? The theory is as primordial and simple as the rows on a vegetable rack: potatoes and carrots - Earth! Lettuce and cabbage - Water! Credulity may be strained by having melons and tomatoes associated with the fire-element. A comment from an astrologically-minded US grower is here helpful:

> Thun has assigned leaf crops to the water element and fruiting plants to fire. Any avid gardener knows that if a fruit plant, such as the tomato, receives too much watering, the green growth is lush while the fruit production is meagre. Conversely, if the plant receives a great deal of sunlight and minimum watering, fruits are large and abundant, even though the plant itself may look as if it won't last another day due to brown and drooping leaves! Function (and activity) seem to be *the* key to the Sidereal Zodiac[7].

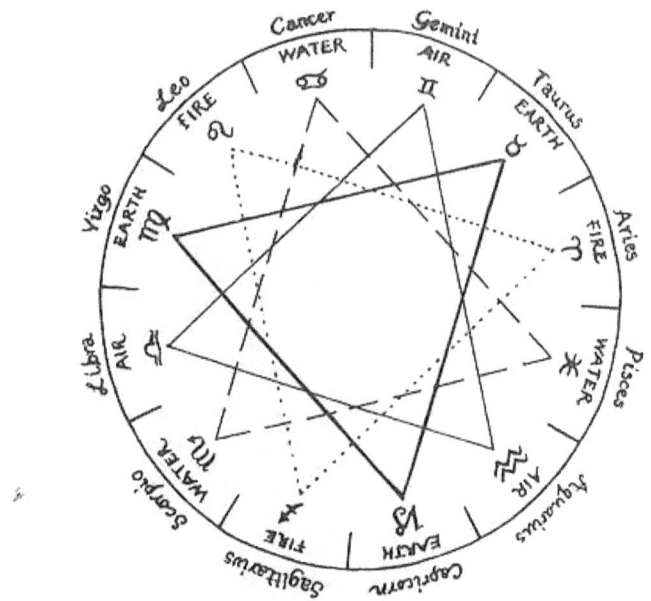

Fig 4.2 The Star-zodiac 'trigons' or sky-triangles of the Four Elements.

---

[7] Letter from Theresa Weed, California, in 'The Astrological Journal' of Spring, 1982 (concerning PBTM).

Sowing crops by Zodiac constellation elements:

Element – Type:	Example /	Constellations
Earth – Root:	Potatoes, carrot /	Bull, Virgin, Goat
Water – Leaf:	Lettuce /	Crab, Scorpion, Fish
Air – Flower:	Cauliflower /	Twins, Balance, Water-bearer
Fire – Fruit/seed:	Beans, barley /	Lion, Archer, Ram

Let us draw a historical analogy. In the second century AD, the astronomer/astrologer Claudius Ptolemy lived in Alexandria, and his magnum opus the Tetrabiblos - a kind of theoretical astrology - contained no hint of the four elements as belonging in the heavens. They were still just the four humours of Greek medicine, describing the composition of the material world. And yet, shortly after, a rather marvelous change happened, something no Chaldean had dreamed of. The qualities of the zodiacal signs started to be described in terms of these four elements. This took a while to gain acceptance, and then it came to seem almost natural, such that astrologers around the world nowadays take it for granted. If it is valid and stands the test of time, will the insight of Maria Thun come to be regarded as such a step?

The theory of Maria Thun was a development from a somewhat obscure[8] teaching Rudolf Steiner had given concerning the 'four ethers.' There were, Steiner said, four types of 'formative force' that worked throughout the realm of nature. These, he said, did not work centrically as the forces known to modern physics, but rather worked via form, and he called these forces 'etheric'.[9,10] This was a reformulation of the ancient doctrine of the four elements, conceiving them as process rather than as substance. Within the Anthroposophical movement, botanical studies of plant morphology by Jochen Bockemuhl have supported the idea that it is helpful to view the stages of plant growth in terms of such 'formative forces' that are linked with the traditional four elements. He has related the stages of leaf, flower and seed formation with water, air and warmth.[11]

The Thun-Heinze Sowing Trials

Let us turn to a series of experiments published as validating the Thun theory, at least for root crops. Statistician Dr Hans Heinze assisted Maria Thun in designing, conducting, writing up and publishing a series of

---

8 Peter Tompkins & Christopher Bird, *Secrets of the Soil,* 1989, p.367.
9 Gunther Wachsmuth, *Etheric Formative Forces,* 1932.
10 Millner & Smart, *The Loom of Creation,* 1975.
11 J. Bockemühl, 'Elements and Ethers: modes of observing the world', in Towards a Phenomenology of the Etheric World, Ed. J. Bockemuhl, Trans. John Davy, Spring Valley NY 1985, pp.1-69, 38.

gardening trials, using potatoes. Three years of potato sowing trials are given in that book, 1963-65.[12]

Other trials with vegetables such as beans and carrots may also have been performed, but these three years of potato trials are the only ones where both sowing dates and final weight yields were given by Thun and Heinze.

Over a twenty-seven day period in April and May, twelve rows of potatoes were sown, one per Moon-zodiac constellation. When they had finished growing, the final weight yields of potato were assessed. This was repeated over three successive years. I have converted the weight-yields into graphical form as shown. To enable such a comparison between successive years, two adjustments to the data were necessary.

Firstly, they were presented as percentages of mean yield, i.e. each year's data was standardized to have the same mean of 100, then a 'season trend' line as shown was subtracted out. The overall yields vary widely from year to year, given in kilograms per hectare.

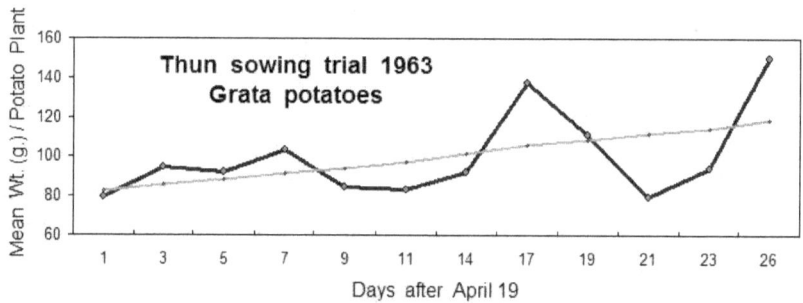

A percentage figure indicates by how much the weight yield has increased when the Moon was in the constellation of the appropriate element at the time of sowing. The element predicted by the theory for potatoes is Earth, and the answer is, 23%.

The nine root-day sowings were above the mean value by this amount: 123±6 [n=9] for Thun's root-day sowings compared to 92±10 [n=27] for all the others, and that is highly statistically significant (t=8). We could say that the root-day sowings were of a one-third greater yield than the other days, on average.[13]

---

12 Thun and Heinze 1979, data from pp. 24-25 for years 1963-5, then pp.34-5 for 1965-70; I am grateful to Dr Manon Haccius for advice over this.

13 There is a converse question, as to how far the yields in Water Moon-signs have been reduced. (reminder: the zodiac sequence is Fire, Earth, Air then Water, the Moon and planets go through in that order). This may be a less interesting issue, as few will wish to reduce their yields by using a lunar calendar, however the overall answer in this case is 12% - a notably smaller effect than appeared in the case of the Earth-sowings excess.

There is or was a widespread belief that Thun has done many more trials over all four 'elements' e.g. beans and lettuce, however I have not been able to find them.

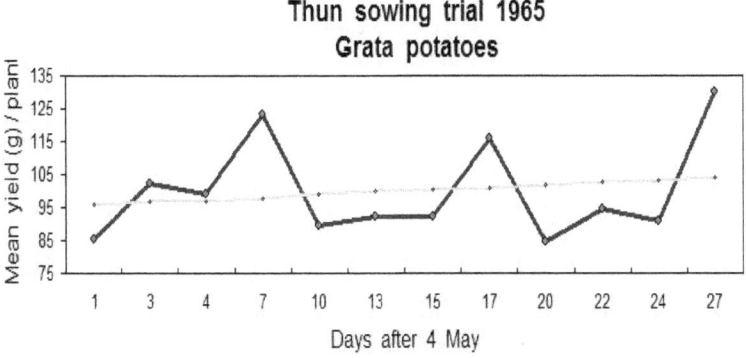

Figs 4.3, 4.4 & 4.5 Potato Moon-zodiac yields, Thun experiments 1963.

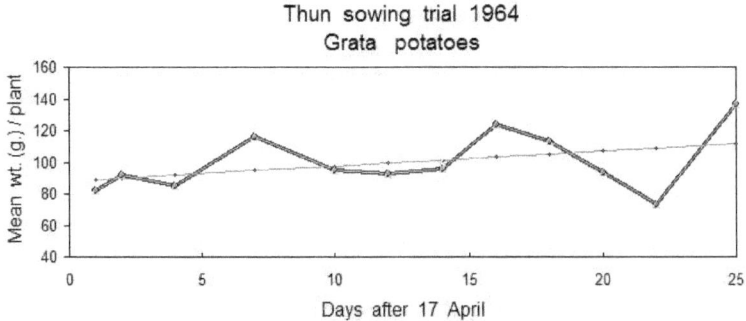

Replications

In Holland, Dr Ulf Abele looked for the influence of a Moon-zodiac rhythm in oats and barley over four years of trials (1970 -'72 and 1974), three of these being part of his doctorate thesis at Giessen on Biodynamic farming. He found that yield increases for barley and oats sown when the Moon was in front of the fire constellations, as predicted for a seed crop, averaged 7 per cent,[14] significant at 1 in 100. Abele found larger yield increases for carrots and radish in earth-sign sowings for trials which he conducted in 1972 and '74, averaging 22% excess, as was highly significant

---

14 Ulf Abele, 1973 (doctoral thesis) and 1975; a brief account in English of the radish experiments of Dr Abele at Darmstadt was given in the 1988 Thun calendar, *Working with the Stars* pp.21-23

(1 in 10,000). Only the 1972 trial was not in itself statistically significant. These results are depicted in figures 5.6 - 5.9. It has been claimed that the extent to which a crop responds to the Moon-Zodiac effect depends largely on the condition of the soil.

Thun emphasizes that a healthy, organic soil which has quality humus is necessary, and that a soil which has been "mineralized" by the use of chemical fertilizers will not respond. As early as 1964 she wrote:

> I came to the conclusion that mineralized soils hardly reacted to these cosmic rhythms and their fine influences, whereas a humus-filled soil of whatever soil type was a good mediator for these forces.[15]

Four years trials of the 'Thun effect' by Abele.

Crop	Data	Mean Y	Trigon predicted (t ha^{-1})	Other rows (t ha^{-1})	t-value
Mean yields and yield deviations					
Barley	1970	3.67	0.29 ± 0.15	−0.10 ± 0.21	2.6 > t_{10}, 0.05
Oats	1971	5.46	0.22 ± 0.45	−0.07 ± 0.15	1.6 > t_{10}, 0.10
Carrot	1972	9.62	1.29 ± 0.60	−0.43 ± 0.61	3.9 > t_{10}, 0.005
Radish	1973	26.25	4.81 ± 1.05	−1.61 ± 2.10	4.7 > t_{10}, 0.001
Combined data with yield deviations normalized to a mean of 100					
Barley + oats	1970, '71		5.7 ± 6.5 (n = 6)	−1.9 ± 4.4 (n = 18)	3.1 > t_{22}, 0.01
Carrot + radish	1972, '74		15.9 ± 5.8 (n = 6)	−5.3 ± 7.3 (n = 18)	6.2 > t_{22}, 10^{-4}

This dependence of the Moon Zodiac effect on soil condition was investigated as part of a doctorate thesis published in Zurich by Ursula Graf.[16] Crops were grown under different organic and conventional agricultural programs and compared. The use of the Moon's "sidereal" position for planting, which belongs to the Biodynamic method of agriculture, was also investigated using "conventional" soil on which chemical fertilizers had been applied.

Graf used essentially the same experimental procedure as Thun, with twelve rows of the same crop sown, one per Moon-constellation, over a

---

15 Maria Thun, 1964, 'Nine years Observation of Cosmic Influences on Annual Plants' Star and Furrow, 22 (reprinted from Lebendige Erde 1963).
16 Ursula Graf, 1977 (doctoral thesis) Darstellung... Zurich Technical College.

The Thun Theory

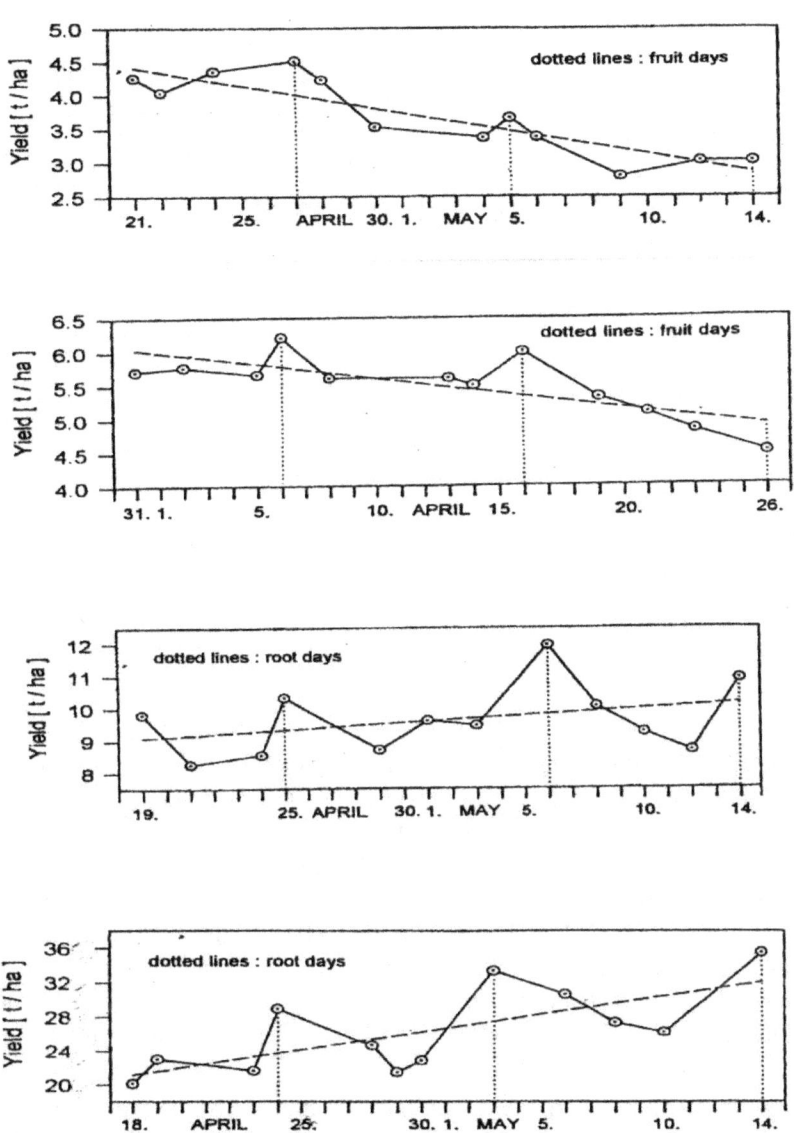

Figs 4.6, 4.7, 4.8, 4.9 Dr Abele's four years of zodiac-sowing trials at Giessen. The graphs plot crop yield against the "sidereal" Moon signs at the time of sowing. The straight lines represent the seasonal trend: Fig 6a Barley 1970, 6b Oats 1971, 6c Carrots 1972, 6d Radish 1974.

"sidereal" month, with final weight yields for each row compared by grouping according to the elements. Potatoes and radishes were used for three successive years, 1973 - 75. While the results were more equivocal than those of Abele,[17],[18] she concluded:

> The soil seems to be a decisive factor in the occurrence of connections between Moon-Zodiac constellations and crop yield.

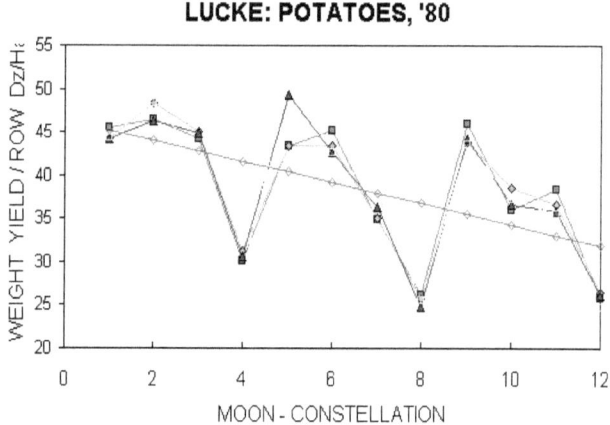

Fig 4.10 The potato Moon-zodiac sowing trials of Lucke at University of Giessen in 1980

A PhD study by Lücke, also at the University of Giessen, investigated root crop yields over the two years 1979-80. His 1980 experiment involved four separate field trials using potatoes, a total of 48 rows (Figure 5.10, Lücke Potatoes 1980). Four trials were conducted in one year at Giessen, to compare the effect of several Biodynamic sprays, each with twelve rows sown over a sidereal month. The graph shows weight yield of potatoes in tonnes per hectare. Earth-trigon (root-day) maxima correspond to the numbers 1, 5 and 9 on the graph. The four trials had very similar mean values, and we could put just a single linear regression line through them (as did Lücke).

This gave a mean of 38.5 decitonnes per hectare, and then after subtracting out the regression line values and grouping the residual values as before, one obtains 5.3±5 [n=12] for root days and -1.8±6 [n=36] for the others, a difference of 18%, a slightly smaller effect than Thun found. This was

---

[17] The lack of statistical significance in Graf's field trials has been pointed out by Spiess (1994, Band3, p.32) and then by Koepf, Schaumann & Haccius (1996, p.204).

[18] The Graf-Keller paper of 1979 plus the Thun-Heinze studies were alluded to by Dubrov (1996), p.117, along with several others, as describing 'very strict standardized experiments ... carried out on plants,' relevant to lunar influence.

significant at 1 in 1000 (t=3.6).[19,20] These replication experiments are further discussed in Chapter Six.

Summarizing, it appears that there are two distinct lunar cycles of primary relevance to plant growth: the Moon's phase, that is its position with regard to the Sun, affects day-to-day growth, whereas the Moon's sidereal position, its position against the stars, can be important when a crop is sown. The extent to which the latter effect manifests depends on soil condition.

Dr Heinze's opinion

Some comments were obtained from Dr Hans Heinze, the statistician who assisted Thun in her experiments, explaining how Thun's background enabled her to made the decisive discovery. I had asked him various questions about how her results were different from Kolisko's, and why it was that Dr Spiess had apparently not replicated the phenomenon.

The words of this old man writing informally in broken English are quoted not least because, over years of endeavor, he was the only person I found in Germany or Switzerland prepared to comment favorably upon the Thun experiments.

Other Biodynamic experts there that I located were seemingly under the influence of Dr Spiess's negative results, and all expressed skepticism, which I found puzzling.

> Maria Thun was born at a small farm where she grew up as a normal peasant-daughter of the time, bringing the farm-products to the market in rustic clothes, till in her twenties she came in contact with anthroposophy through her friend and husband, an anthroposophical artist and teacher. But she was living with the stars and nature, already in her youth by the special endowment of her peasant-father in this line, having every day the constellations of the stars in her mind, teaching also about this topic.
>
> As you see, this is just the opposite, of what you pretend to be a good condition to get objective results - not to know anything about the constellations of the stars at the day you are sowing [I had suggested that an experiment should ideally be done 'blind', such that the experimenter would not know what celestial events were going on.]. But, in contrast to what I thought to observe, accompanying Lilli Kolosko at her work, Maria Thun is not living

---

19 J. Lücke, 'Untersuchungen...,' 1982, pp. 71, 74.
20 The shorter account, 'Mond-Trigon-Wirkungen: Eine statistiche Auswertung,' Lebendige Erde November 1998, 478-483, reviewed experimental investigations of the Thun effect since the 1970s, and a translation appeared in Harvest, Biodynamic journal of the Seasons, New Zealand, May 1999.

with her will-power in her work, to bring the constellations into manifestation. She is only living with them, as the mother is living with her children. Thus she is meaning, that everybody should have the same results, because she is not doing anything special to get them, but living with them. Thus there arises the question - what is objective in this situation?

These remarks give food for thought concerning what kind of experimental attitude is appropriate, and why Kolisko's work has not generally been replicable in contrast with that of Thun. Dr Heinze added a comment upon lunar phase, not incompatible with the views of Thun:

In these experiments we came to the result, that the days before full-moon show well a better harvest, but the plants tended to fungus-attack and worse quality.

Concerning Thun's work with milk and butter production, Dr Heinze wrote that over a four-year period, in four months out of every year, with two selected goats, Thun had an:

... astonishing result, of 70% more butter in the warmth-trigon against the three others, and 89% more butter in the constellation moon in Erdferne against in Erdnahe, apogee against perigee.

Regrettably, German Anthroposophists use the term 'constellation' for just about anything happening in the heavens. Here it is referring to the monthly apogee and perigee positions. 'Warmth-trigon' alludes to the three fire-element constellations in the zodiac. I did not confirm that any such result was present for the apogee-perigee cycle, in the two months' data which Thun published (see next chapter).

One must regret that Thun's book about milk production and other aspects of life on a Biodynamic farm[21] did not remain in print for long. It gave a holistic perspective of life on such a farm, being pleasantly illustrated with goats, cheese, ducks etc, and indicated how the sidereal patterns give a work-schedule for structuring the course of the week. It gave daily yields of milk and butter production for August and September 1980 on which Heinze's comments were based. Let us hope that it will be translated into English. In the meantime, here are some further comments of Dr Heinze, from another letter of his:

---

21 Maria Thun, *Milch und Milchverarbeitung* 1985.

Fig 4.11 The wise woman: Frau Maria Thun making cheese (From her 'Milch und Milchverarbeitung'). The pole signifies the Axis Mundi.

Concerning the Milk-Book, because some of our scientists could not imagine that a person without an academic degree was able to produce scientific, reliable, experimental results, the Milk Booklet is written as a peasant woman for peasant women. But nevertheless, the results are exact. During the months of the test, she worked, besides her daily work as gardener and in the household, until every midnight to handle the milk of the day and rose in the morning at 6 o'clock.

I asked Dr Heinze as to why merely eight years of root crop experiments had been written up, while one hears stories of twenty-five or thirty years' of experiments on the matter by Thun (assisted by Walter and Matthias Thun, her son and husband). He replied that there had indeed been further experiments conducted with positive results over many years, and "Also farmers and gardeners have good results." One would like to see the records of what has been found in such German experiments, both positive and negative, collated, if they are still accessible, plus the testimonies of farmers who are prepared to comment upon whether the Thun calendar has helped them. After all, this is where it matters. One aim of Thun's years of potato sowing trials was to investigate seed quality. Each year, 'eyes' were cut out of the potatoes and sown, instead of the whole potato, in the same Moon-constellation. Heinze commented:

> These years were enough for us, especially for the question, whether we were able to keep the breeding quality of the potatoes with this method, and we succeeded very well over these years to keep it.

There is one person I knew who has travelled quite a bit in Switzerland and Germany meeting Biodynamic farmers, Freya Shikorr. I asked her what attitude she there had found towards the Thun calendar, and she said that, while the Biodynamic 'experts' were mainly skeptical, the farmers were generally pleased with it and found that it worked. This skepticism derived partly from the experimental work of Dr Spiess, which came out in the 1980s, and partly from the fact that Steiner's lectures on agriculture given in the 1920s made no direct allusion to such sidereal rhythms.

As regards the question of repeatability, here is what Dr Heinze had to say (quoting from two separate letters):

> Dr Spiess working with his test at the Dottenfelder Hof Bad Vilbel very exactly, but not thinking about the moon and with 150 test-plots has no results at all. A straight line is formed by his harvest of the test plots through all moon-constellations.
>
> There are people who have the same results as Maria Thun, but there are others, carefully working, who have no results at all - we could not find an explanation for this fact in the state of the soil, climate, etc.
>
> My personal opinion is, that the reason is to be found in the quality of the person. This influence is known from a teacher entering a class or from every artist, one has the capacity and the other not, it is also known of some plant-seed breeders, and also in Java, in Indonesia, where I lived for some years. One man was sowing the rice-plants for the whole "dessa" village, because, as one said there, he had a warm hand.

This discovery certainly is making difficulties for the acknowledgement, but on the other hand it is important, that we have to educate such capacities, to breed new plants for nourishing mankind when the present ones are worn out, by only exploiting them, without any knowledge, how they were bred in the mysteries of Persia, 5,000 years ago from wild plants, giving the marvelous gold color to the grains, which the grasses do not have, and are losing now, when treated with artificials.[22]

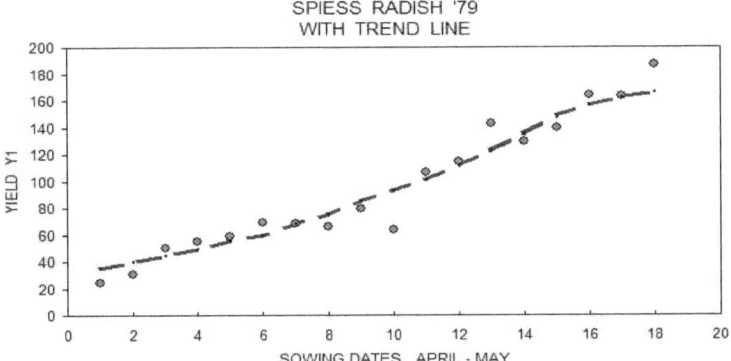

Fig 4.12 Spiess radish trial of April-May 1979, with seasonal-trend line.

Fig 4.13 The same, 1980: with and without added fertiliser.

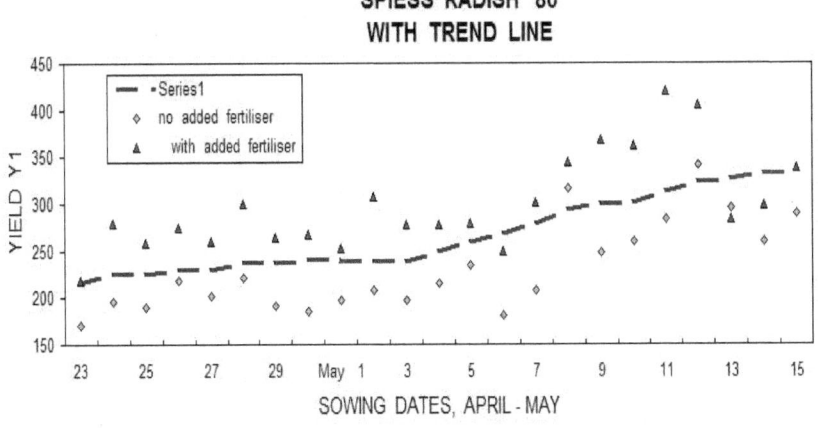

---

22 Decades earlier, Hans Heinze had conducted a little-appreciated study of formative forces working in ice-crystals: H. Heinze, 'Einiges uber kunstliche Eisblumen', in 'Kristalle,' Stuttgart 1930 25-31.

This brings into the foreground an important point also being raised in modern physics, whereby the experimenter is inexorably linked with the experiment. It does seem that a person's attitude can affect the growth of plants she or he is tending. Studies on this matter are hard to come by, as if it were a matter quite beyond the pale for a botanist to investigate.

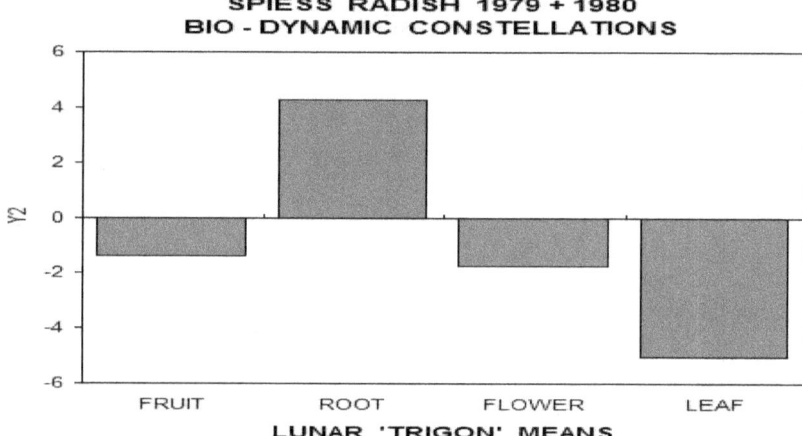

Fig 4.14 Two sets of Spiess radish data combined, 1979 & 1980, after subtracting out trend lines, grouped by 'trigon' element at sowing date.

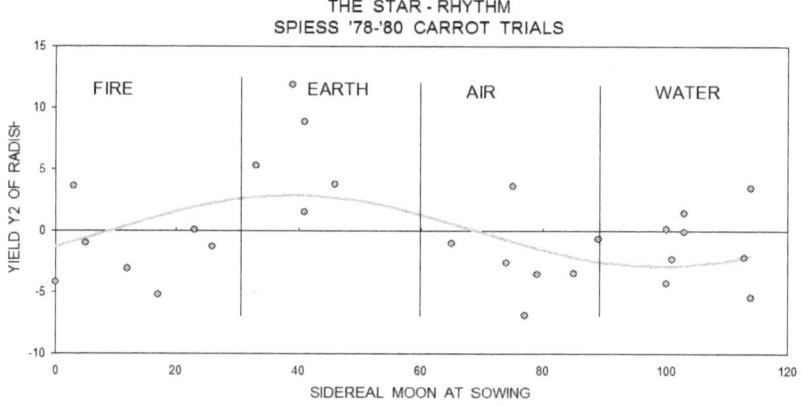

Fig 4.15 Spiess carrots over two years, plotted by lunar position at sowing date over 120° of celestial lunar longitude, to show the Four Elements, with best-fit sinewave.

Nonetheless, it is widely believed and would indicate the sensitivity of plants to external factors. The viewpoint here taken differs from that of Dr

Heinze, for instance I found that Spiess did obtain significant positive results in his trials, albeit of small amplitude.

I'm not accepting a certain subjectivism implied in his viewpoint, that a mere different attitude or presence of the experimenter can cause the phenomenon to vanish. It would not be feasible to develop a scientific approach to the phenomenon, if it were the case that a mere negative attitude could erase the effect!

The Spiess Challenge

In 1984, Spiess and Schaumann replied to an article by a German Biodynamic farmer[23] casting doubt upon his belief in the 'Thun effect.' Their experiments, they explained, conducted at the Dottenfelder Hof farm near Frankfurt, had failed to display any effect of the elements of the constellations against which the Moon stood at sowing date, upon final yield, in contrast with Herr Lust's experience.[24] This exchange took place in the German Biodynamic journal 'Lebendige Erde'.

Rumors of this major series of experiments, scientifically and rigorously designed, with careful controls, where the advice of Thun herself had been sought for the experimental setup, spread. When the present writer in the early 1980s now and then approached Dr Herbert Koepf, at Emerson College, who was in charge of the Biodynamic agriculture course, on the matter of evidence for the BD calendar, he would mutter darkly about these trials not having found anything.

A further Spiess article in 'Lebendige Erde' in 1987 reviewed the whole subject and reaffirmed the negative results of himself and others.[25] In 1990 Spiess published two reports, comprising an analysis of some of his experiments, in the organic gardening/farming journal, 'Biological Agriculture and Horticulture'.[26]

In 1994 there appeared Dr Spiess's magnum opus, a two-volume 600-page statement.[27] It endorsed certain traditions at variance to the Biodynamic

---

23 Volkmar Lust, Lebendige Erde, November 1984, pp.257-261 'Eininge Praxiserfahrungen mit den "Thun'schen Aussaattagen."'
24 W. Schaumann and H. Spiess, 'Mitteilung in Zisammenhang mit dem Bericht von V. Lust', op. cit. p.261-262. Hartmut Spiess works at the Institute for Biodynamic Research, Bad-Vielbel, Germany.
25 Hartmut Spiess, 'Zur Frage der Wirksamkeit kosmischer Rhythmen und Konstellationen,' Lebendige Erde, 1987, 6, pp.305-315.
26 Spiess, *Biological Agriculture and Horticulture* 1990, 7, pp.165-178: 'Chronobiological Investigations of Crops Grown under Biodynamic Management. 1. Experiments with Seeding Dates to Ascertain the Effects of Lunar Rhythms on the Growth of Winter Rye,' followed by 'Chronobiological Investigations ... on the Growth of Little Radish.'
27 H. Spiess (1994) 'Chronobiologische Untersuchungen mit besonderer Berucksichtigung lunarer Rhythmen im biologische-dynamischen Pflanzenbau,' Darmstadt; 2 vols, I 'Band 3'

farming calendars, such as that below-ground crops such as potatoes were best sown at New Moon and above-ground crops best sown by Full Moon; it averred that an apogee-perigee effect was consistently present in the data, which was the reverse of that implied by the BD calendar, whereby crops sown at perigee grew best; and claimed that the Thun effect, which forms the backbone of BD calendars, was absent from the data-sets. The two volumes were published by the German 'Forschungsring', the prestigious research-circle of Biodynamic farmers located in Darmstadt. One presumes that they endorsed it.[28]

This fairly strong fourfold negation, expressed over a period of one decade, had repercussions seemingly wherever the subject of Biodynamic calendars was raised. For example, the 1993 'Lunar Organic Gardener' of Llewellyn, US, commented that Spiess 'could not find significant data to support the specific planting-day theory',[29] while an article in a Dutch journal discussed the crisis of confidence that has arisen amongst BD calendar users, as a result of the results from Dottenfelder Hof farm.[30] A New Zealand book about Biodynamics commented upon the inability of European researchers to confirm the Thun-effect.[31] In 1993 Koepf published a booklet 'Research in Biodynamic Agriculture' where Spiess's comments upon his trials were the only evidence discussed for celestial influence as used in this context.[32] His previous book on BD agriculture, a few years earlier, had equivocated over the relevance of the sidereal month for agriculture and notably failed to endorse the Thun theory. The 1996 edition of this textbook was likewise in tune with the wave of skepticism over the calendar sweeping over German BD experts.[33]

---

of 258 pp., and II 'Band 4 - beschreibung der einzelergebnisse' of 319 pp. (the experimental data)

28 Earlier, Thun's own researches had been 'Thanks to the financial support of the Research Ring for Biodynamic Agriculture, and of former friends, as well as the co-operation made possible with the Institute for Plant Propagation and Cultivation at the University of Giessen, under the direction of Prof. Boguslawski...' (Thun calendar 1974 (English translation), p.48.

29 Llewellyn 1993, p.90.

30 Hans van den Bosch, 'De Zaaikalender werkt heel anders,' Onkruid 79, Netherlands, March/April 1991, pp.144-149.

31 *Biodynamics - New directions for Farming and Gardening in New Zealand*, New Zealand BD Assoc., Auckland, 1989, p.131.

32 Herbert Koepf, *Research in Biodynamic Agriculture: Methods and Results*, Biodynamic Farming & Gardening Association Inc USA 1993, 74 pp. Koepf's *The Biodynamic Farm* 1989, NY, endorsed sowing by the waxing Moon and nothing else: 'sowing seed in damp earth during the waxing Moon furthers the growth of plants.' A statistically insignificant table of results by Graf was cited, with reference to the Thun effect (p.116, yields of radish in a growth chamber), which the text did not endorse.

33 The 1996 third edition of the textbook on Biodynamics by Koepf, Patterson & Schumann devoted several pages to discussing the Spiess research, hardly mentioning any other, and averred quite incorrectly that Ulf Abele's four years of trials with grain and root crops at

An article co-authored by the present writer with a German statistician was published in 'Lebendige Erde', reviewing evidence pro- and con- the 'trigon' effect. The journal published it with extensive comments both before and after as to why farmers should not believe in the sidereal rhythms! Initially there was an editorial, then came a five-page interview with Spiess on the subject. A letter appeared in the following issue endorsing this view. The co-author was fed up with this highly partisan approach, but eventually a second article was translated into German and submitted. It reviewed Spiess's work and refuted its claim not to have found any sidereal rhythms in crop-yield.[34] 'Lebendige Erde' eventually published it, much abridged.

The Thun effect was fairly consistently present as a low-amplitude effect in the years of Spiess's data. Our approach to the data-analysis differed from Spiess's as we used a moving-average to allow for the seasonal trend, where sets contained very large seasonal trends which were far bigger than any monthly trends.

He used certain mathematical curves, which did not well model the seasonal effects. He did radish and carrot-sowing trials over three years, 1978-80, and we found that, combining these all together, his yields from root-day sowings were in excess at the highly significant level of 1 in 1000 - even though, overall, the mean excess in weight yield was a mere 8%.[35]

Three years of Spiess carrot data was analyzed in the same way: the three sets of data were first converted to the same mean of 100 so that deviations from the mean were percentages, then moving-average trend lines were put through these three data sets, subtracting out which gave 27 data-points; they have here been plotted on a 120° dial, of sidereal Moon-signs, and a best-fit waveform put through them.

This is a very low-amplitude Thun-effect, a mere 3%, but it may be of theoretical interest that it is present. One can see that the waveform here plotted has this 3% amplitude, because the mean value is 100. An article by this writer and and Dr Staudenmaier, reluctantly published in the prestigious German Biodynamic journal Lebendige Erde, argued that this data did support the basic Thun-hypothesis. See Appendix 2 for more details.

Dr Spiess has argued that an error has been made by Biodynamic farmers in adopting Thun's theory, and that the indications given by Dr Steiner to

---

Zurich had given negative results. This writer and Dr Staudenmaier set the record straight (see next ref.)

34 N.K. & Gerhard Staudenmaier, 'Mond-Trigon-Wirkung: eine statistiche Auswertung' *Lebendige Erde*, November 1998, pp.478-483: translated in *Harvests*, the NZ Biodynamic journal, as 'Maria Thun's Trigons, What have Other Investigators found?' Winter 1999, pp.13-17. Spiess critique by N.K. & G.S. in: Mond in Tierkreis: anders rechnen - andere Ergebnisse, Lebendige Erde, Jan 2001 pp.48-9 (much abridged).

35 N.K. & Staudenmaier, 'Evidence of Lunar-Sidereal Rhythms in Crop Yield: a Review,' Biological Agriculture and Horticulture, 2001, 19, pp.247-261, 257.

farmers concern merely the synodic cycle.[36] The Dottenfelderhof farm near Frankfurt, where the trials were performed, is in a quite highly industrialized area of Germany and suffers from acid rain, which could here be relevant. A historic approach is really necessary for resolving this question, which now follows. This may help to ascertain to what extent the quest for a Biodynamic calendar is getting anywhere.

Chronology: A Century of Endeavor

The oldest account of a lunar agricultural calendar in Europe is the 'Works and Days' of Hesiod, composed in the time of Homer;[37] different days of the (lunar) month were sacred to various deities, giving them different qualities.

From that archaic, magical view of the ancient Greeks, there developed the Roman traditions of certain days or phases of the month that were crucial, described in Pliny's Historia Naturalis (chiefly in Volume VII) composed in about AD 60. Since then, these ancient traditions have splintered into diverse traditions of folk-belief.

An outline of the historic sequence of events relevant to Biodynamic calendars follows. The chronology touches upon synchronous events involving the rediscovery of the star-zodiac of antiquity, as used by the ancient Chaldeans and Egyptians, in the first half of the twentieth century,[38] which have been placed afterwards.

1912: In Berlin, Rudolf Steiner recommended a new approach to the experience of the twelve constellations of the zodiac, and presented his Kalendar 1912/13. It gave positions for sidereal divisions of the twelve constellations, and 'Not only was the moon's quarter given, but also the constellation occupied by the moon each day.'[39]

1924: Dr Steiner's Agriculture lectures led to the founding of Biodynamic farming movement. We have already quoted his comment there about 'How the stars work in plants'. They also included the categorical statement that; "we shall never understand plant life unless we bear in mind that everything

---

36 H.Spiess, 1990, p.166; a similar view was expressed by Koepf (1989). For a review of the debate, see David Wright, 'Maria Thun's Moon-days, Fact or Fiction?' *Harvests* (New Zealand Biodynamics) 2004, 57, 18- 20.
37 Hesiod, *Works and Days* Loeb Classical Library, Heinemann, 1977.
38 See Michael Baigent, *From the Omens of Babylon* 1993; *Astronomy Before the Telescope*, Ed. C. Walker 1996, Ch.3.
39 In 'Concerning Moon-Harmonics in Plant Growth' (*Mercury Star Journal*, Winter 1978) I claimed that in 1976 Robert Powell had argued (MSJ, 'On the Divisions of the Firmament' Easter and Summer 1976) that Rudolf Steiner's indications concerning where the boundaries between the constellations should be drawn corresponded more closely to the Babylonian equal-interval sidereal zodiac than to the divisions nowadays embodied in the BD sowing calendar.

which happens on the Earth is but a reflection of what is taking place in the cosmos."[40]

1932: Steiner died in 1926. A book by Guenther Wachsmuth about the four ethers appeared in 1932. These were envisaged as formative processes that worked throughout the realm of nature. While making many suggestions about them, it did not link these 'four ethers' as Steiner had described them with the Moon's passage around the zodiac.

1935: An article entitled 'The Significance for Seed-germination of the Passage of the Moon through the constellations of the Zodiac' appeared in a British Biodynamic magazine.[41] Its authoress Maria Hachez alluded to ongoing experiments showing the effect of lunar phase on seed germination, clearly referring to the work of Kolisko though without mentioning her name, and continued:

> ...in addition to the effects of the lunar phases, the influences of Cosmic forces having their origin in the passage of the Moon from one constellation of the Zodiac to another, could also be investigated. These further experiments have been carried out during the years 1930-1935, at the Observatory of the Mathematical-Astronomical Section of the Goetheanum, under the direction of Dr E.Vreede, and with the help of Herr Joachim Schultz. As they were followed up these Cosmic forces of the constellations could be discovered active in the building up of all matter, in the nature of the structural substances, and in all that brings about the finer differentiations of form, quality and taste.

The experimental procedure was outlined:

> Carefully selected seed of various plants was sown at intervals of two to three days, in long experimental beds in the open...We recommend others who may wish to try similar experiments, to sow the seed at the same hour in the morning or evening and at intervals of 2-3 days.

Four decades elapsed before anyone in the UK conducted such experiments.

1939: Madame Kolisko, an Austrian who had come to England, apparently in some degree because of rows within the Anthroposophical Society, wrote her pioneering work, Agriculture for Tomorrow. It dealt with lunar phase rhythms in the realm of nature. Graphs indicated maximal seed germination on the days before Full Moon. Years of results (that have proved largely

---

40 Rudolf Steiner's *Agriculture,* a series of lectures, 1927, 1991 Edn. Kimberton, PA, p.27. The sole monthly cycle to which these alluded was the synodic (p.23), however they also referred to the 'ascending period' of Saturn (p.26, note 10 p.266). It seems that, by analogy, this has led to the Biodynamic adoption of the tropical month: Vreede 1936, Ch.2, ref 23.
41 Maria Hachez, 1935.

unrepeatable) were shown, of maximal crop yields for Full Moon sowings.[42]

1948: Franz Rulni produced the first Biodynamic calendar, due to last into the late 1970s. It contained advice on the mating of livestock and made predictions concerning gender of offspring. The Rulni calendar introduced the 'obsi' and 'nidsi' Moons, from the Emmenthal valley in Switzerland. These are equivalent to what later became known as the ascending and descending Moons. For example, a 1978 Rulni calendar advised for the month of June:

> With the beginning of 'nidsi' Moon, prepared compost should be spread on all mown meadows and pastures ... Good planting-out time.

For November:

> The nidsi moon period should be used for winter cultivation for seed beds for rye and wheat in order to enhance good root formation and tillering... Moon becomes nidsi: use next fortnight for cultivating fallow fields which will then keep clean and weed-free until spring.

The fortnight of 'nidsi' is thus more important in the B.D. calendar than the 'obsi' period, which is the other half of the (tropical) month. Of a May Full Moon in 1978, Rulni's calendar advised:

> Full and New Moon times are detrimental owing to the after-working of the eclipses in April. Especially inhibiting due to the node, Full Moon and Perigee occurring so close together. Once these are over and the ground has been well-prepared get on with all important sowings.

1956: Maria Thun in Germany had her idea about the four ethers/elements. While Maria Hachez had described supposed effects of individual constellations, Thun perceived the primary effect as emanating from the zodiacal elements. Thun started sowing rows of radish, then other annual crops, and at first it appeared to her that root growth began when the Moon reached its descending node. Then her attention shifted to the waning Moon:

> Over and over again, the waning moon proved itself to be the main factor, which is reinforced when the descending day forces are active at the same time. Thus it was frequently noticed that plants which were transplanted in the afternoon at the time of the waning moon only needed watering at transplanting time and then grew without check.

---

42 Replications of this effect in the 1930s by others were however reported: Kolisko, 1933, and the June 1931 issue of the UK Biodynamics journal, pp.55-56.

It is possible that Thun meant to refer to what Biodynamic farmers call "the descending moon" (if so, she was certainly not the last to be confused over these issues). From observing her radishes, she came to realize that the plant should be viewed as a fourfold being:

> I suddenly found that I had discovered a law of life that governs plant growth. Had the first breeders of our cultivated plants an insight into this law? The discovery of this fourfold nature of the plant was an undreamed-of reward for the work of many years.[43]

She envisaged these four stages as energized by the sidereal lunar orbit, and then realized, she said, that a lecture by Dr Wachsmuth had made the same claim. We may compare this with Robert Powell's version of events:

> It was Guenther Wachsmuth who first postulated (in a lecture) that since a certain element predominated in each of the four types, then the signs of the zodiac are likely to exert a corresponding influence. For example, the watery element is predominant in cabbages, therefore the growth of cabbages should be enhanced by the signs of Cancer, Scorpio and Pisces. Thun proceeded to test this hypothesis.[44]

She then investigated this pattern with a variety of crops: carrots, parsnips and scorzonera were tested as representatives of root-types; lettuce, spinach, corn salad, cress and a few brassicas as leaf types; zinnia, snapdragon and aster as flower types; beans, peas, cucumber and some tomatoes as fruit-seed types. Cucumbers sown on leaf days showed lush leaf growth but a reluctance to flower. Leo, she concluded, was especially beneficial for seed formation. Of what constituted good soil, she observed:

> It is evident that it is a soil which though good cultivation with adequate dressings of compost has the power of receiving cosmic influences and transmuting them into processes necessary for growth, of being able to bring star-forces to activity within the plant whereby new qualitative relationships are brought about, producing food fit for man.

Once Thun had decided to use the constellations, she turned to the 'Sternkalendar' of the Anthroposophists which used unequal constellation

---

43 Thun op. cit. 1963, p.5. Her discovery is described in Thun, 'Cosmic working in soil and plant of sidereal Moon-Rhythms,' *Mercury Star Journal* Summer 1978, pp.42-51. Being familiar with Franz Rulni's calendar, she sowed several different batches of radish seed, and 'the shapes of the root-bulbs were so varied that this phenomenon gave me no peace.' She sowed radish batches daily, and the next year she let some go to seed, and 'Only then could one see that we had to do with 4 distinct types.' A few years later in 1956 she heard Dr Wachmuth's lecture about the four ethers in relation to the Zodiac.
44 Robert Powell, 'Lunar Calendar for Farmers and Gardeners', *Mercury Star Journal*, Summer 1977, p.55. This journal appeared quarterly for the five years 1975-9.

boundaries. These had been derived from boundaries fixed by the International Astronomical Union in 1928. The I.A.U. had there ruled that thirteen constellations lay across the ecliptic. Anthroposophists modified these boundaries somewhat, for reasons not easy to come by, and changed the number of constellations back to twelve. These matters will be the subject of a later chapter.

1963: Thun published her first lunar gardening calendar. There were two main differences between Franz Rulni's calendar and the Thun calendar which became its successor: Thun removed the synodic or phase cycle, inserting in its place the four-element sidereal rhythm. Both calendars were built upon the four lunar months (Ch. 3), but differed in their choice thereof. Rulni placed more emphasis upon the adverse effects of eclipses, as shown by the above quotation, while Thun only avoided the day of an eclipse. Rulni set great store on the usually favorable influences of the monthly Moon/Saturn oppositions, while in the 1970s the Thun calendar used no planetary aspects; they had however become incorporated into her calendar in the 1980s. Over seven years from 1963 to 1970, Thun performed systematic trials chiefly using potatoes and these were later written up jointly with statistician Hans Heinze. Her experiments, conducted over decades[45], were made on field plots in her large garden.

1973: The 1970s were a decade when independent corroboration of the Thun effect started to appear. In 1973 Dr Ulf Abele presented his research findings from growing barley, as part of a PhD on Biodynamic farming methods, finding that yields were increased significantly on 'seed-day' sowings. Later in 1977, Graf reported some fairly positive results with root crops and different soils. Experiments in Britain (organized by this writer) were first reported in 1977.[46]

1978: The US Kimberton Hills Agricultural Calendar began, which has grown to a circulation of 7000 annual sales, with recommendations very similar to the Thun calendar.[47] It gives the various celestial events, conjunctions and oppositions, lunar and planetary each month which the Thun calendar does not.

---

45 There is no experimental write-up of any Thun vegetable-growing trials, except those performed with Dr Heinze, a fact which (I was advised) contributes to the mood of skepticism that is prevalent in Germany. Brief outlines of some of her experiments are given in *Hinweise aus der Konstellationsforschung,* translated into English as 'Results from the BD Sowing & Planting Calendar' 2003.
46 Nk, 'Zodiac Rhythms in Plant Growth – Potatoes' *Mercury Star Journal*, Summer 1977.
47 The Kimberton Hills Agricultural Calendar was renamed 'Stella Natura' and is available from: P.O. Box 550, Kimberton, PA 19442, USA.

1980: Two new cycles appeared in the Thun calendar: that of apogee/perigee, then in 1981 the nodal cycle.[48] These transferred over from Rulni's calendar, but I don't know how Rulni came by them as their use was not traditional. Up till then the Thun calendar had comprised solely the sidereal and tropical cycles, these having the same 27.3 day period (Ch.3). Two components of the Rulni calendar which Thun never adopted were the phase cycle and the monthly Moon-Saturn oppositions. Planetary trine aspects made their appearance in 1982. The British lunar-gardening guide 'Planting by the Moon' appears, by Simon Best & N.K., and uses the equal-interval star-zodiac.

1984: Dr Spiess began publishing reports describing his results from the late '70s and early 1980s. He carried out sequential sowing experiments with winter rye and radish, reporting synodic and perigee effects but finding no sidereal pattern in final yield.

1992: The first articles by wood-forestry expert Dr Ernst Zurcher appear, which endorse the Kolisko Moon-phase findings for germination and growth.

1993: The American Llewellyn 'Moon Sign Book', which enjoys a larger annual sale than any other publication in this area, and has been going since 1904, changed its name to the Lunar Organic Gardener, and featured major discussions about the Biodynamic sowing calendar. It referred chiefly to the Kimberton Hills calendar, and summarized parts of Maria Thun's book, 'Work on the Land and the Constellations' in a fairly sympathetic manner. Concerning the irreconcilable differences in theory and practice, it struck an optimistic note:

> Even though the two systems differ so drastically (one system's fruitful dates may be the other system's barren dates) the reasons and basic tenets are the same.[49]

On this, we may demur.

2001: A comprehensive review of evidence for the 'Thun effect' was published in Biological Agriculture and Horticulture. It refuted Spiess's claim.[50]

The new millennium saw a series of articles by Dr Ernst Zurcher working at the prestigious ETH (where Einstein had worked) in Zurich, confirming

---

48 Thun's 1977 publication *Work on the Land and the Constellations* had described the effect of these two cycles, indeed her very first essay on the subject (1963) described how she experienced the growth-inhibiting effect of perigee upon radish.
49 'Biodynamic Gardening' in Llewellyn's *Lunar Organic Gardener* 1993, pp.216-220, Foulsham's. I wrote to Llewellyn pointing out the incompatibility of the two systems, and suggesting a series of experimental trials to ascertain which one worked better, but received no reply.
50 Op cit (34).

the presence of synodic monthly and daily lunar rhythms in tree metabolism.

2002: Careful research from the Hiscia Klinik (for cancer research) in Switzerland is published, showing a zodiac-element Moon-rhythm in a changing shape of mistletoe berries.[51]

2006: The BBC filmed a successful Moon-gardening experimental trial by a student at the Royal Horticultural College at Kew, Thea Pitcher, using this writer's lunar calendar.[52]

2009: A slim new Thun booklet 'When Wine Tastes Best' receives near-rapturous endorsement from top British wine-tasters, supermarket chains and even newspaper columnists: based on a polarity between the elements of Fire and Earth, where warmth-days lead to an experience of the best taste, while root-days are the worst.

A UK Biodynamics textbook, 'Cosmos Earth and Nutrition' by Richard Thornton Smith, endorses use of the equal-interval star-zodiac in agriculture.

2010: The London International Wine Fair annual meeting promotes Thun's new wine-taste fire-trigon theory.[53]

2012: Death of Maria Thun

While the Anthroposophists were taking a new interest in the constellations, the original star-zodiac was being discovered. Anthroposophists have taken no interest in this phenomenon, being prone to an anti-astrology bias. The cuneiform code of the ancient tablets was cracked around 1890, and scholars realized that they were looking at the birth of the zodiac itself (This story gets told in Chapter 6). In the 1970s, both Robert Powell and the present writer came to advocate use of this

---

51 Baumgartner et al., 'Form...Mistelberen,' 2002, Elemente der Naturwissenchaft, 77 (2) pp. 2-15; followed by their 2003 article, 79 (2) pp.2-21. English translations: 'Shape changes of ripening mistletoe berries', Archetype No. 9 (Ed. David Heaf), September 2003; 'Mistletoe berry shapes and the zodiac,' No. 10, September 2004.

52 Thea Pitcher (the BBC's DVD, 'A Year at Kew, Series Three' 2007, Episode 9, online if you google her name) grew various sets of lettuce, sweet peas, onions, swiss chard and leeks at Kew, and assured me (and the BBC) that she had obtained around 30% yield increases from sowing at the 'proper' time. It was done as a gardening project for her horticultural course. Thea would not accept any suggestion of mine that her expectations might have affected the results. I held back completing the present opus for a couple of years, hoping that she would write up her successful lunar-gardening experiment performed at the Kew Royal Horticultural College, filmed by the BBC and using my calendar, but ... she didn't.

53 *When Wine Tastes Best, A biodynamic calendar for wine drinkers* 2010 Maria & Matthias Thun, Floris Books, 2009. The Thuns claimed not to have discovered anything: '...this recommendation doesn't come from Maria Thun. It's actually the wine industry itself which has discovered the tendency for many wines to be at their best on fruit and flower days...'

equal-interval Sidereal zodiac for the Biodynamic schedule.[54],[55] Powell was then editor of the Mercury Star Journal, in which the first articles were printed. The Biodynamic journal Star and Furrow printed my proposal, although its advisors rejected the notion.[56]

In the 1980s the mantle of siderealism descended on the booklet 'Planting by the Moon' by Best and Kollerstrom. These authors brought out a yearly Moon-calendar through the 1980s using sidereal ingresses computed by Neil Michelson at AFA San Diego, and it appears that no other sidereal journal functioned over that period. That manual integrated the insights of Kolisko (phase cycle) and Thun (sidereal), while rejecting that of Rulni (ascending-descending cycle). In 1982 Neil Michelson, director of the American Federation of Astrology publications, published a sidereal ephemeris,[57] and this was used for the PBTM calendar.

---

54 Powell op. cit. (44), p.57: "The reason why Biodynamic farmers use the constellations is primarily historical, i.e. in the first half of this century the constellations, handed down from Greek astronomy, were the only sidereal divisions generally known. It is only in the second half of the twentieth century that widespread knowledge of the signs, originating from Babylonian astronomy, has been acquired. Even now the signs are to a large extent unknown, but thanks to the painstaking work of a number of science historians, especially Otto Neugebuer, the highly sophisticated Babylonian astronomy of the last centuries B.C. has been reconstructed in some detail."
'The Sidereal Zodiac by Powell and Treadgold appeared in 1979, and was reprinted in 1985 by the American Federation of Astrologers.
55 N.K., 'Moon Harmonics in Plant Growth' Mercury Star Journal, Winter 1978, p.150.
56 Star and Furrow, 1982, 59, pp.23-3, with comments by George Corrin and John Soper; correspondence reviewed by J.T. Burns, *Cosmic Influences on Humans, Animals and Plants,...* 1997, pp.68-9.
57 Neil Michelson, *The American Sidereal Ephemeris,* 1976 to 2000, A.F.A. San Diego, California.

5. Design of a Time Experiment

Give me the ways of wandering stars to know,
The depths of heaven above, and earth below,
Teach me the various labours of the Moon.
                               - Virgil, Georgics, II, 475.

The first crop-yield experiments designed by the author, were to test the conclusion of Kolisko, that yields varied with lunar phase,[199] a view expressed in her book, *Agriculture of Tomorrow*, a pioneering work on the practical aspects of Biodynamic farming.

These trials had sowing dates extended over several lunar months. Whereas 'Thun-experiments' reported on the Continent have consisted of rows sown over at most one sidereal month, here the sowings extended over two or three lunar months.[200] The longer sowing period meant sowing a larger number of rows, which led to an improved statistical treatment of the results. As others had noted, the effect as described by Kolisko was not evident in the results.

In these experiments, one plants a sequence of rows of the same vegetable over a month or two, and grows them over similar conditions, thereby providing a tabula rasa on which Mother Nature, the Cosmos or whatever, can reveal any time-rhythms that may be working.

It is preferable for such trials to extend over a longer period than one sidereal month, since, if the periodicities involved are of a monthly nature, then they will hardly be evident if the experiment only has twenty-seven

---

199 Kolisko (1936, 1978) outlined sowing trials 1926-34, concluding that sowing two days prior to the Full Moon gave optimal crop yields. Others reported quite similar results: in 1931, the UK BD Journal reported three years of investigation by the German J. Voegele of oats, wheat and rye, showing that sowing just before Full Moon gave optimal yields, while those just before New Moon gave the worst, measuring yields in cwt/acre. Another experimental station in 'middle Germany' found that lettuce sown just after the New Moon grew better than that sown just before. A Mr Conradt found that his maize sown just before Full Moon grew better than that before New Moon (reported in Anthroposophical Agricultural Foundation, Notes & Comments, June 1931, p.56). Negative results from such trials were claimed by Mather in 1942 (Appendix 1).

200 Thun and Heinze 1979: a collection of articles published in 'Lebendige Erde' between 1963 & 1972, mostly experiments with potatoes. See also Thun 1977. For a discussion by Thun of Kolisko's work and her attitude towards the 'Full Moon effect', see her 1980 Calendar, pp.17-20. Some more recent experiments of hers (also of Mr Schwartz) reported in the updated 'Work on the Land and the Constellations' 2nd Edn. 1991, pp.49-51, were reproduced in the Kimberton Hills Calendar of 1995.

days as the sowing period. We will see in subsequent chapters how this works out, and what the implications are of designing an experiment over a longer sowing season.

The method of sowing twelve rows over one sidereal month used by Thun amounts to putting a question to Nature, to which she is permitted only a single yes/no reply.

Is the effect working or not? As there exist four primary lunar-monthly cycles (Chapter Two), one would rather see experiments that are capable of discerning which of these (if any) may be operative, instead of assuming that we are looking at one of them. Dr Spiess claimed that the apogee/perigee cycle was showing up in his experimental trials, and the Bishop experiments in Wales sometimes showed pronounced growth-inhibiting effects of the lunar nodes. To examine such factors a sowing period of at least two months is preferable.

Three university studies have tested and confirmed the Thun model, and together with the experiments here reported bring the number of persons who have published confirmation of the theory up to a good half-dozen.[201,202,203,204] The experiments published in this work remain the only systematic British investigation of the theory.

They suggest that sufficient tests have been carried out to show that the lunar effect is real and needs to be taken seriously. The trials here described were carried out on a limited budget and, although I hope that the basic methodology is sound, there is plenty of room for further expansion and refinement of this work.

Having said that, one should note that a textbook on Biodynamics by Koepf *et. al.* has taken rather an opposite view: its readers will gain the impression that the above-mentioned university studies have been inconclusive or negative, failing to achieve statistical significance. The aim of the present work is not (I hope) to try and convince the reader of a particular viewpoint, but rather to present as fully as possible the relevant issues, in an area that most educated persons regard as merely a joke, and which is rather riven with contrasting and incompatible opinions.

The means whereby the cosmos, nature and the human being interact are mysterious and sensitive and knowledge of this area remains at a rudimentary stage. The most thorough and systematic research project to-date has been that by Hartmut Spiess, which has supposedly failed to demonstrate the sidereal lunar effect. The Spiess work has been a stimulus

---

201 Ulf Abele 1973 and 1975.
202 Ursula Graf 1977 and 1979.
203 Colin Bishop, 1977.
204 J. Lücke 1982. A confirmation of the Thun effect using potatoes in 1969 by a Mr Schwartz was published in the 1983 Thun Calendar, p.16.

for the collection and publishing of these essays. It may precipitate a crisis amongst the Biodynamic community, as far as the use of calendars is concerned.

Considerations of Quality

The outcome of a properly conducted experiment should be public knowledge, i.e. knowledge which can be shared. If, for example, one tasted carrots from several sowing dates, and declared on that basis that a certain sowing time in the month was optimal, few would be be convinced. As a personal experience it might be vivid, but the method would be unduly subjective. It might convince one person, viz. the experimenter. The expectations of the person, or the 'prejudice' as a critic might say, is able to influence the results in such a case.

A taste experiment could be done by several persons at the end of the month after all rows had been harvested. Each taster could score the taste of each sample, on a scale of for example one to four. A properly conducted experiment has to involve some quantitative scale, or the knowledge gained remains private. For example, I enjoy cutting open a red cabbage to inspect the pattern inside, believing that one can discern the healthy form of an organically-grown cabbage or the deforming of proper growth due to chemical fertiliser-stimulated growth, which leaves a chaotic structure. Were a few red cabbages from each row of an experiment to be cut and photographed, or otherwise preserved, and then compared at the end of the experiment, with two observers asked to grade each one for 'degree of form,' then those derived from leaf-day sowings might score better, according to the Thun theory.

Organic growing is supposed to bring about an improvement in food quality. Jack Temple, who was the Here's Health organic gardening correspondent, performed such a quality test in the spring of 1982, using the Thun sowing calendar:

> Every time a leaf-sowing date turned up I sowed a row of lettuce and a row of radishes, and every time a root-sowing day featured in her calendar we also sowed both lettuce and radish again. Then, when the directors and students from the Henry Doubleday Research Association dropped in on their yearly visit we put the trial to the test.

Initially, Jack's visitors were shocked by the outlandish notion, so that his reputation with them appeared to be at stake.

> However, all that disappeared as I put my knife through two radishes. One was juicy and the other had the texture of cotton wool. The juicy one had been sown on the correct day, a root-day, and the pithy radish had been sown on a leaf-day. That was not

all: the juicy radish was ten weeks old while the pithy one was only eight weeks old. Both radishes were also tasted for quality and flavor...subsequently further sowings were tested. Each time we had the same result. Radishes sown on leaf-days were pithy and radishes sown on root-days were firm and fleshy. The lettuce trials did not produce quite such clear evidence, but lettuce sown on leaf days were slower to bolt than those sown on root days.[205]

It was thereby shown that root-day radish had better texture and the leaf-day lettuce were slower to bolt, in front of some fairly critical witnesses.

Thus, one should by no means take the view that a 'scientific' experiment is one which looks only at weight. We are concerned with a calendar system applicable to organic growers, people who have made a personal sacrifice to reduce yields substantially by renouncing chemical fertilizers, for an improvement in quality.

What do we mean by quality? Let's turn by way of reply to an experience of Sherry Wildfeuer, which led a decade later to her starting up America's main Biodynamic calendar, that of Kimberton Hills.[206]

Genesis of the Kimberton Calendar

The testimony of Sherry Wildfeuer was written up in the 1986 Kimberton Calendar, and is worth quoting at length because it shows so well the issues involved in designing a crop yield time-experiment:

> In 1970 the quest for training in Biodynamic agriculture led me to the garden of the Goetheanum in Dornach, Switzerland, which is the world centre for the Anthroposophical movement arising out of Rudolf Steiner's work. Part of my job was to carry out the replication of an experiment which Maria Thun had done in Germany showing the effect on plants of the Moon's passage through the zodiac. The same crops were grown side by side in four patches and treated identically except for the timing of their culture according to the element of the constellation in which the Moon stood.
>
> Every few days, as the Moon progressed, I would be busy: first preparing the soil, then sowing, cultivating, weeding, and later spraying the Biodynamic silica preparation. The care was normal,

---

205 Jack Temple, 'Checking the Value of planting by the Zodiac' *Here's Health,* November 1982, pp.144-5. Temple's book 'Gardening without Chemicals' (1986) is compiled from his 'Here's Health' articles as their organic gardening correspondent, and contains several accounts of how he found use of the Thun calendar improved quality and yield.

206 The Kimberton Hills community and its lunar calendar is well described in *Secrets of the Soil* by Bird and Tompkins, 1989, Chapter three. The Calendar has been renamed as *Stella Natura,* available from PO Box 550, Kimberton, PA 19442.

albeit more intensive than usual, particularly with regard to the frequent spraying. This was intended to intensify and make more visible the different qualitative effects of the light.

In reporting on my observations I must point out that this occurred in the very early days of my gardening career, and there may well have been more to see had I been more astute. As it was, although I looked regularly, as the plants were growing I could notice no difference amongst the patches. What was striking was rather the health and vigor common to them all, which I naturally attributed to their intensive care.

In autumn I was sent out with wooden crates and a fork to harvest the roots. I laid out a crate by each patch and began to dig carrots, beginning with a Fire patch. As I expected, there was a good crop from each. But when I came to the Earth patch I was struck by the beauty and ideal form of the carrots. Lovers of plants will know well the experience of being filled with delight and appreciation on contemplating a truly beautiful specimen, and wanting to call others to come and share the wonder. This feeling grew and grew as one after the other carrot proved to exemplify the "ideal carrot" of my imagination. And to top it off, I had filled the crate before finishing the harvest, and had to go fetch a second box to hold what was obviously a larger yield from the same-sized patch. This done, I returned to dig up the beets, and found, to my great excitement, the phenomenon repeated. Again the Earth patch produced aesthetically beautiful roots, and required a trip to the shed for a second box to hold the harvest, which had not been necessary from the other patches.

Soon after I was given four paper bags and asked to gather the calendula seeds. This was not a pleasant job because the flowers were mostly black and mushy, and the seeds were not fully ripe. But in the Fire patch this was completely different. There the seeds were fully ripe, so much so that it was a problem to collect them, for many had fallen to the ground and already germinated! That impression of the Calendula seeds went in deep. It meant that I had living proof that the Moon was indeed a gate to qualitatively different forces in the heavens. As a gardener, the scope of relevant concerns had now extended not only theoretically but quite practically to the world of stars.

It was a great privilege to have taken part in such an experiment. Perhaps never again will I have an opportunity to perceive the effects of the stars on plants so clearly, for in the fullness and diversity of life's involvements experimentation is not the most urgent priority. Yet for myself a conviction was born at that time which did not require re-proving, but rather the kind of confirmation which comes from working with a principle and seeing it prove itself true again and again in myriad subtle ways.

One is taken aback to hear that this experiment was never written up. Though it would appear to have been very thorough, indeed a very model of how such an experiment should be performed, no record of it exists outside Ms Wildfeuer's account.

I wrote to Ms Wildfeuer for her opinion regarding the experience of verification in the US, and her reply was:

The people I know who work with the Calendar and have strong confirming experiences are all practicing farmers and gardeners, not scientists. There are certainly many 'stories,' but these lend themselves more to personal sharing than to print... It is a mystery that these influences show themselves so clearly to some, and remain hidden from others.

An American grower wrote of Thun's yearly calendar, especially its weather predictions, that:

It is in this context that Maria Thun relates earthly weather with the alignments of the stars and gives the world a planting calendar that has proved itself to its severest critics - at least here in the Alleghany Mountains of Northwestern Pennsylvania.[207]

While fully respecting such experience, the stark fact remains that other growers feel with equal conviction that their experience testifies to a different system, based upon the traditional tropical zodiac - of a comparable age to the Thun calendar, if not older. An article by the present writer in the US 'Biodynamics' journal stressed the importance of conducting some time-experiments in the New World, only not performed using the primitive method of sowing two sets, one at the 'right' and the other the 'wrong' time.[208]

The book, 'Biodynamics', published in New Zealand, expresses a pessimistic view about Thun's experimental work: that it had "not been able to be duplicated in trials in Europe to-date (although experience in New Zealand indicates that the effect of the constellations can be discerned)."[209] One would like more of an account of how New Zealand farmers were experiencing a confirmation of the sidereal pattern. One is not complaining

---

207 Biodynamics' (quarterly magazine), Kimberton, Pennsylvania, Winter 1991-92, p.37, letter by L.A. Rotheraine.
208 N.K. Testing the Lunar Calendar, *Biodynamics*, (KimbertonUSA) Winter 1993 pp44-48.
209 Biodynamics 1989 New Zealand, p.131.

that five published reports of positive results by the present writer[210] should have passed unnoticed, but it seems odd that doctorate theses by Abele, Graf and Lucke which did in some measure confirm the sidereal cycle should also be passed over by the antipodean commentator. Rumors of unrepeatability would seem to be gaining momentum. It may be that we stand at a turning-point. Biodynamic calendars cannot continue to develop with the sole systematic set of experiments supporting their claims having been performed in the 1960s.

There have been UK tests (other than those organized by the present writer) of the Thun hypothesis: by Jack Temple as mentioned, by the Centre for Alternative Technology in Wales, and presently by 'Gardening Which?' magazine.

I wish to argue that these time-experiments are fun to perform & that the experience of communing with Mother Nature through such, preferably without too much of a preconceived idea of what the results are likely to be, is inherently rewarding.

Does it matter, for example, when seed is harvested, for growing future vegetables? Should one collect it on a fruit/seed day, or would a root-day be better in the case of carrot seed? Thun has expressed differing views on this matter,[211] and an experiment would help.

The Experimenter as Subject

A theme still highly censored is (for want of a better expression) the person-plant relationship. Do persons have green fingers, causing their plants to thrive? There is little published on this vitally important area of person-plant interaction, which remains beyond the pale for biology departments. The attitude of the experimenter may well be relevant to the result obtained.

For this reason, one should doubt experiments where merely two different plots are grown, one sown on what is supposedly the 'right' time and the other on the 'wrong' time, as giving undue scope for the attitude of the experimenter affecting the results. In her 'Planetary Planting', Louise Riotte describes her garden in these terms, so visitors can be pleasantly surprised by how much better is one side of her garden, planted according to a lunar calendar, then the other. The first principle of an experiment is, that a sufficient number of rows should be sown. The gardener tending the plants should be impartial.

---

210 NK: 1977, 1978, 1981, 1984 and 1985.
211 The former practice seems to be advocated for carrot-seed in Thun's 'Working with the Stars' 2002 (p.66), from a 3-year carrot experiment, though she had hitherto advocated the latter: Maria Thun, Gardening for Life – The Biodynamic Way, 2000, p.22.

Karen Herms is a German lady who now lives in Britain, and who has visited the farm where Spiess was performing his experiments, and that where Thun conducted hers. The farm where Dr Spiess was working (Bad Vielbad) was, she told me, surrounded by industrial development and situated in one of the worst acid rain areas in Germany, while Thun's farm (near Marburg) lies in a very tranquil rustic area surrounded by woodland and steep hills. Persons informed on these matters are hard to come by. We need to understand better the kind of influences which interfere with or enhance the effect; the person/plant relationship on the one hand, and then the presence of pollutants in the biosphere, etc.

An essay of Goethe's had the title, 'The Experiment as Mediator between Subject and Object.' Today's scientific training involves a rejection of any such concept, but we may be inexorably driven to acknowledge that, somehow, the subjective being of the experimenter is relevant. One wishes that more experiments could be performed on what appears to be a strictly censored area: for example, if a person suffering from chronic depression holds a glass of water, and then seeds are watered with it, will they germinate less well? Can this be shown reliably? As an experiment this is easy enough to perform. We should comprehend how sensitive germinating seeds are to various kinds of influence and observer attitude if we are to be capable of understanding the matter of celestial influence. For example, the orientation of seeds to magnetic North does enhance their germination.[212]

Let us hope that in the future, scientists will not be so frightened of discussing their attitude towards an experiment. One hears stories about how the apparatus simply would not work while so-and-so was in the lab., and this may be what is happening in these replications. If Hermetic principles are to be formulated linking Earth and sky, they are not going to apply with unfailing regularity as do the laws governing the downward fall of an object. Living things are more unpredictable than the non-living. If we are to respect the integrity and being of the creature, and try to become aware of the cycles of time to which it is attuned, then an attitude more of patient attention as Goethe recommended is called for.

Up the Garden Path

There were some Californian studies in the 1950s by a minister, published as 'The Power of Prayer on Plants',[213] stimulated by the parapsychologist J.B. Rhine at Duke University. In one experiment, two jars of water were prepared, and one was brought into a 'prayer circle' and passed around. Then, various pairs of pots with soil and seeds were prepared, each watered

---

212 Bruce Cumming in *Geo-Cosmic Relations,* 1990, p.50; Dubrov, *The Geomagnetic Field and Life,* 1978, p. 216-7.
213 Rev Franklin Loehr, *The Power of Prayer in Plants* NY 1959.

with one or other of the jars. Quite large differences in growth would sometimes appear. Sadly, negative thoughts directed towards the seedlings produced even larger effects:

> The usual prayer-plant experiments, with only positive-prayer seedlings to compare with the neutral non-prayer ones, produces a relatively small difference in growth, running usually from 5 to 50 percent. But an effective negation experiment will produce differences of several hundred percent and more.[214]

One would like to see such experiments performed in school biology lessons. They have a twofold relevance to the present study: in demonstrating the sensitivity of plants to external factors, which is in a general way supportive of our argument; and in undermining arguments based upon dividing a garden plot into two sections, one of which is grown in accord with the gardener's belief-system and the other not ...

A biochemist at McGill University in Canada found that if plants were watered with weak saline solutions that had been held by healers, then these plants grew better than plants watered with the same solution but not so held.[215] What did this imply? The biochemist next took two identical bottles of water, and gave one to a hospitalized depressed patient to hold, while the other was held by a 'normally happy' man, for half an hour. Plants watered from the bottle held by the normal person grew and thrived, while those exposed to the water from the other one languished. Such an experiment confirms how sensitive and receptive is the germinating seed. One is perplexed that a phenomenon so fundamental, and of such general relevance to the human condition, has received so little by way of recognition.

In 'The Secret Life of Plants,' authors Tompkins and Bird described how the attitude of the grower could affect the plant grown, then they returned to this theme in the sequel, 'Secrets of the Soil,' which described how Russian Alla Kudryashova was able to 'energize' grain before it was sown in fields in Puschino near Moscow. On one-hundred acre plots, beet seeds were supposed to have increased their weight yield considerably after she had 'touched' the sacks containing them. These experiments were written up in 1987, and apparently approved by Russian agricultural authorities.[216]

A series of such trials were performed by two biologists at the University of London.[217] They germinated cress seeds and grew them for a week. They found that in six out of seven trials, using a person they regarded as a skilled

---

214 Ibid, p.49.
215 Bernard Grad, (McGill University, Montreal) *Int. Jnl. Parapsychology*, 'A Telekinetic Effect upon plant growth,' 1964, 6, pp.473-498.
216 Tompkins and Bird, Secrets of the Soil, 1991 p.325.
217 A.M. Scofield & R.D. Hodges, 'Demonstration of a healing effect in the laboratory using a simple plant model' Journal of the Society for Psychical Research 1991 57 p.312-343.

healer, significant differences in the growth of seedlings were found between the healing and non-healing group. They cited eight previous studies which had found positive results, and six that had reported negative results - i.e. that human will or prayer made no difference to seed germination. They concluded 'Although we consider that the results obtained do support the contention that a healing ability exists, it is unlikely that any laboratory model will ever satisfy those who do not wish to know.' On the other hand, three years of field trials by Serena Rodney-Dougal found that on a farm seeds that had been prayed over by a healer, holding them in his hands (which she called 'enhanced' seeds) did not grow better:

> The results do not favor the hypothesis of greater germination and growth, but there is a measure of support for better health." The healer had "no noticeable effect upon germination.[218]

Such prayer-experiments involve a control experiment, comparing two sets of growth. There is no harm in unusual concepts being proposed, as long as a commitment to verification is present. For example, one sometimes hears it said that seeds are improved if stored in a pyramid, but evidence has not supported this.[219]

Rudolf Haushka claimed that seeds germinated in a sealed container (near a Full Moon, in early spring) altered in mass, and in both of his books he showed graphs over years of work apparently demonstrating this.[220],[221]

I repeated this several times with a high-precision balance and six silicone-sealed glass vessels, three with seeds germinating in and three without, and took weight measurements to one order of magnitude higher accuracy than he had used, but found no weight change. I saw no flaw in the experimental procedure as Haushka described it, and have complete belief in his integrity, but I merely claim (as others have done) that it was a result which I could not replicate.[222]

Louis Kervran wrote several books claiming that living things and especially germinating seeds could transmute elements, one of which was

---

218 S.M. Rodney-Dougal & J. Solfin, 'Field Study of Enhancement Effect on Lettuce Seeds: their Germination rate, Growth and Health,' *Jnl. of Soc. Psychical Research*, July 2002, 66.
219 Margaret Baker, *Gardener's Magic and Folklore*, 1978, p.31-2: the Cheops Pyramid Company of St Louis, Missouri was claiming that their model pyramids would 'stimulate plant growth' and 'sprout healthier seeds quicker.' This was checked at the University of Guelph, Ontario, by Prof Herman Tiessen where various different seeds were sprouted inside and outside various different pyramids, but no differences were found.
220 Rudolf Hauschka, *The Nature of Substance*, 1966, pp.16-20.
221 Rudolf Hauschka, Nutrition, 1967 pp.132-137.
222 Stephan Baumgartner's 'Hauschka's Weight Experiments: Weight Variations in Germinating Plants in a Closed System,' (Dornach, 1992, in German) found slight weight-changes when seeds germinated in sealed containers, varying with the phases of the Moon (Burns 1997, p.100).

translated into English. That sixties classic 'The Secret Life of Plants' by Tompkins and Bird endorsed Kervran's claims about elemental transmutation, however their more recent opus took a more cautious line on the subject.

I looked into this using germinating seedlings. The procedure is complicated, and involves reducing the seedlings to ash by incineration and then dissolving this in acid to form a solution, then the metal ions could be assayed photometrically. I had the opportunity to do this at a technical college at Ewell, but I did not fully explain to the authorities what I was up to, as it would have sounded rather peculiar.

A Government chemist 'R.M.' guided me carefully through the procedure and helped with the microassay. I was rather expecting the experiment to work, as Kervran had said. But, as my experimental procedure gradually improved, the differences between, say, calcium present in the control (ungerminated seeds) and that in the seedlings progressively decreased, and it finally became evident that the differences had no significance beyond experimental error.[223]

Others have also reported a failure to replicate the Kervran effect. Here, one's attitude appears not to affect the phenomenon, and it would appear to-date that germinating seeds do not transmute elements.

More recently, as interest has developed in water-quality, and whether water can be somehow 'energised,' re-informed or whatever, charming 'flowforms have been developed which have water cascading down through lemniscates curves. They are a delight to watch, but to they 'improve' the water? In 2004, batches of lettuce were grown using 'flowform' water and compared with lettuce grown using ordinary tap-water. The book 'Energising Water' by three authors published in 2010[224] averred that they grew thirty percent more by weight. I visited Emerson College in Sussex where these experiments were done (and where the flowform was invented) and was assured of this result – but also, that two further attempts at replicating this effect in different locations had not shown any such result. That information should really have been included in the book. If it doesn't repeat, it isn't science.

People are interested in quality-testing, using form. One thinks here of the huge interest in Mr Emoto's water-crystal research, where his Japanese laboratory photographed water-crystals from clear and polluted water-

---

[223] I was advised that the seedlings had to be grown in sunlight, and then that they had to be germinated in mineral water, not distilled water, but I still obtained the negative results.
[224] J. Schwuchow, John Wilkes and Ian Trousdell, *Energising Water, Flowform technology and the Power of Nature* 2010, p.82.

sources.[225] We appreciate the simplicity of this experimental approach, looking at what is so close to life itself: water. Can it be refreshed and revitalized? But, after a decade of his pictures being widely admired, no-one other than himself it seems has managed to produce these images. He photographs ice crystals as they are just being formed. Do these images represent a fusion of Science and Art, a laboratory-produced expression of formative forces? Repeatability is necessary if the subject is to become a science, and let us hope some progress will here be made. Such a qualitative approach has to be vital for a development of the topics here discussed.

A Rival Show

A vigorous US tradition over lunar gardening that has endured a little longer than the Thun theory, is diametrically opposed to it.[226] Its tradition of empirical research into the subject grew up at a similar point in time to the Biodynamic and Thun investigations in Europe.

In 1945, the Research Committee of the Planetary Planting Society of New Orleans was organized. A decade earlier, Charles R. Hook Sr., M.A.F.A. became convinced of the method: "To carry out this research" he reported in the 1942 American Federation of Astrologers, "I planted two gardens every year over a long period of time - one at the proper time and one at the wrong time. In each instance the same soil, fertilizers and treatment were used, the culture and care of the plants was the same, and only a pathway divided the two gardens. Seeds were saved from each garden over a long period of years, and these were replanted in their respective gardens. The difference between the two was marked." [227] Both seed fertility as well as gross yield were said to depend on whether the Moon at sowing was in a fertile sign (water signs of the tropical zodiac) or a barren sign (fire signs in particular). Leo he regarded as especially barren:

---

225 Masaru Emoto, *The Message from Water* 1999. HADO publishers, Japan 2000. shows with exquisite photography how crystal formative forces will respond to human attitude, music, ambient conditions and general pollution. Translated into various languages, this inspiring book represents a major breakthrough. For an account of how to make the crystal-patterns, and Emoto's philosophy of 'Hado' concerning the subtle energy, see Lawrence Ellyard, 'The Spirit of Water, The hidden Message for all of us', 2007.
226 In 1939 a Dr Timmins of Chicago published 'Planting by the Moon,' affirming that the results of gardening trials by himself and others in America endorsed using the Tropical zodiac. I haven't seen a copy. He generally evaluated the signs separately, but viewed the water signs as the most fertile. The Llewellyn Moon Sign Book has been published annually since 1906.
227 Mrs J.H. Watts, 'Moon sign and Phase Planting Experiments' American Federation of Astrologers 1946 Yearbook

Three plantings of corn in the zodiacal sign of Leo for three years, caused germination tests to show a fall from 90% to less than 30%, with a very poor yield.

Is Leo the 'wilting sign' as US planters aver? Biodynamic farmers on the contrary regard the Leo constellation as the very best one for seed potency. These signs and constellations are not the same, but they overlap one day every month!

Mr Hook was urged by Llewellyn George, a leading US astrologer and publisher of the annual 'Moon Sign Book' since 1905, to retail the zodiac seeds he had been breeding, and so 'Moon Sign Growth Seed' became a thriving business enterprise, with Mr Hook broadcasting a regular Saturday program on the local radio station in North Carolina. Sometimes, Mr Hook came across the problem that gardeners, on hearing his radio program, would seek to implement his advice using an astronomical almanack which used the constellations, and he would have to explain why these were inappropriate.

'Moon Sign Garden Clubs' began to spring up across the US, dedicated to showing the truth of these astrological principles. Twelve persons interested in astrology and gardening should be elected as officers... The formula was always the same, that above-ground crops were to be sown in a waxing Moon and below-ground crops in a waning Moon, and that the three water-signs were fertile. 'Thoroughbred seed' could be produced in three years, they asserted.

For Llewellyn, Leo is 'the most barren sign, and used only for destroying weeds and other noxious growths,'[228] whereas Thun finds the Lion-constellation optimal: "To grow good seed, choose Lion days which are particularly suitable. These are designated by Fruit-seed days..."[229] In her 1996 manual she returned to the subject: "For many years the best time for harvesting cereals had always been found to be when the Moon was in the constellation of the Lion ... As regards yield, there was nothing comparable to that from seed which had been harvested when the Moon was in the Lion."[230] Here, the best time to harvest refers, not to the yield, but to the germinating power as of the cereal as will appear the year after.

The Llewellyn and Thun views on Leo are thus especially opposed and incompatible. They express two different interpretations of the fire-element. The former views fire as dry and destructive, while the latter is more alchemical, shall we say, of fire as the last stage of a development earth-water-air-fire, and relating to the maturation of seeds in late summer. The

---

228 Charles R. Hook, *The Lunar Farmer* AFA 1947 Yearbook, p.29. I am grateful to Geoffrey Dean for providing these two sources.
229 M. Thun, *Working with the Stars,* 1992 p. 27.
230 M. Thun, 'Working with the Stars,' 1996 p.9

constellation of the Lion BD farmers take as 138°- 173°, from zero Aries, while the tropical sign of Leo is 120°- 150°; there is a twelve-degree overlap, i.e. about one day per month! "One should cultivate a mood of ironic reflection on such days" I commented (in 'Planting by the Moon' 1999).

This view is expressed in the popular US work by Louise Riotte, 'Planetary Planting' - which contains no word about the planets. Some comparative set of trials should be arranged: one cannot have two mutually incompatible lunar gardening traditions, both claiming experimental support extending over decades. One tradition must surely fade away.

The extensive experiments reported by the Planetary Planting Society and by the annual Llewellyn Moon Sign Book have involved two sowings, one at a 'right' time and one on a 'wrong' time, so that the gardener's expectations can plausibly influence the growth, in whatever way one wants to suppose this can take place. Recent research into the extent to which plants can be influenced in this manner leaves open the possibility that this may account for such different results:

> We astrologers can do much to educate the public when and how to plant in harmony with Nature's laws. We can organize Garden Clubs and instruct the members and audience how to use an authentic Moon Sign Book, not almanacks computed by astronomers...[231]

These texts show just the same duality as is found in Biodynamic treatises, though in reverse, as it were. Neither of these references are advocated by the present work. The Moon's position within the tropical zodiac is not a phenomenon of Nature, but is an artifact resulting from a sequence of mathematical computations.

How can a germinating seed respond to something that can only be discerned from computed ephemeris tables? No human being could discern the Moon's position in the Tropical zodiac, so should one expect more from a plant? The following was the diplomatic conclusion advocated by Best and Kollerstrom in 1982:

> Incidentally, we are not implying that astrologers should be using the sidereal zodiac for their work. Rather, it seems that different phenomena may be attuned to different systems. The tropical zodiac is a moving zodiac, moving around 1° every 72 years against the fixed stars, and this evolving system may be valid for man.
>
> However, plants are simpler in their organization than man, and have a far longer history, two factors which seem to have inclined

---

231 Llewellyn's 1993 Lunar Organic Gardener, MN USA, p.25.

their response to the Moon in terms of the more primal and unchanging sidereal zodiac. The position of the Moon in the tropical zodiac requires calculation, whereas its position in the sidereal zodiac can be observed in the sky.

Although plants do not respond to the tropical zodiac positions, it may well be that man's being is more in tune with the special mathematical treatment of time and space on which the tropical system depends.[232]

The Concept of Significance

Traditional lunar almanacs have a claim to be based upon years of experience and sometimes upon experiment. Here for example is an antipodean astrologer who has given the usual views for when to sow (fire signs are barren, water are fertile, and waxing or waning moons are advised for above or below-ground crops respectively):

> Some readers may like to try an experiment to test the validity of the Moon's magical influence upon gardening. From experience, I know you will be amazed at the results. Follow the lunar timetables and rules at the end of this section...

Four boxes are required, and, the author confidently predicts:

> The difference between the strong, luxuriant plants from box number one and the weak, spindly ones from box number four will surprise you, especially when it is remembered that all the seeds came from the same packet and shared the same conditions.[233]

These were the principles that his father had used before him, whereby only one-quarter of the zodiac is really fertile. Large yield differences were perceived between sowings on the 'right' and 'wrong' days, as indeed Louise Riotte averred in her book based on comparable principles.[234]

Experiments involving at least a dozen rows and preferably more are required to resolve the deep and perplexing difficulty in this situation. We have reviewed investigations showing that plants are susceptible to the attitude of the experimenter, though this remains a censored and taboo zone for modern biology. It is therefore feasible that an expectation or hope of prolific growth of one set of vegetables could tend to produce such. Sowing a larger number of rows in one experiment, where the sower does not know

---

232 Best & Kollerstrom, *Planting by the Moon* 1980, Foulsham, p.36.
233 Richart Sterling, *The Complete Book of Astrology*, Australian Consolidated Press Sydney 1980 p.31, 'Moon magic in your garden.'
234 Louise Riotte, 1975, 1982.

which are meant to be the optimal times, will help to remove this effect based upon the expectation of the subject.

The next chapter will describe some more experiments on the basis of which Planting by the Moon was launched in the 1980s. If their results are true and repeatable, then our very perception of the universe we inhabit is altered.

In these experiments we end up by merely comparing numbers which represent two groups. In fact, we compare six numbers: mean, standard deviation and number of rows for two groups of sowings, for one of which a yield increase has been predicted by the theory. A ratio is obtained, which then translates into 'significance'. That ratio will tell us, what should be evident from inspecting the graph. If we see a plainly meaningful pattern, then the test will emerge as "significant".

The concept of significance relates to what was discussed earlier as public knowledge. Persons unfamiliar with the experiment can yet apprehend the concept of statistical significance. There are procedures associated with such a concept, such as that the prediction must have been clearly specified in advance. There are some more obscure notions such as that the distribution must be 'normal' for the test of significance (called a 't-test') to be applied. Sample size is crucial and if possible one should have more than three rows to a group, three being the minimum required for obtaining a standard deviation. The significance derives from the yield excess as compared with the standard error of the group. Standard error measures the overall scatter of the data. As a rough guide, for statistical significance, a predicted excess must be greater then the standard error value.

However, statistical significance is far from being the only issue in these experiments. Usually one has first to separate some estimate of the seasonal trend, as varies through the months, from any lunar periodicities which will be cycling round within each month. The simplest approach is to put a best-fit straight line through the data, or alternately as we shall see a 'moving average' may be preferable. This is a reason for conducting the experiment over a period longer than one sidereal month: it enables a better estimate of the seasonal trend to be made. Dr Spiess published three years of radish data, for one of which a seven hundred percent yield increase developed during the course of the month. In our BAH article (Biological Agriculture and Horticulture) Dr Staudenmaier and I expressed doubt as to whether data containing such a steep seasonal trend could really be analyzed for monthly rhythms. All of Thun's experiments somehow maintained a fairly level trend line over their 27 days of sowing, whereas mine have usually required some arithmetic procedure for separating seasonal trend.

The next chapter requires stamina, because it analyses data using alternative zodiac reference frameworks, as employed in different calendar-systems. The aim here is to compare their utility. Earlier publications broke the data down into four mean element-yields, which, using the different zodiacal frameworks, bewildered everyone, including myself. Instead, here the data has simply been divided into two groups, the one element for which an excess has been predicted, and then all the others. The chapter concerns only one monthly cycle, the sidereal. The last chapter will consider whether more than one cycle or pattern can be discerned in the data, and will endeavor to use wave-equations to describe the effects observed. That will enable a final statement upon the matter.

Let us conclude with some words from a review of gardening folklore:

> At present all that can be said with certainty is that the cosmos seems clearly to influence the gardener's world. Yet traditional beliefs still stand in disarray, with contradictions and obscurities perplexing enough to daunt most ordinary gardeners. Perhaps further carefully controlled experiments may lift the mysteries from the realm of folklore and provide practical formulae for all to use. When this comes about, gardeners could at last have come to working terms with beliefs of great antiquity.[235]

It is heartening to realize that this great transformation is in our lifetime taking place, not least from the work of Zurcher and his colleagues in Zurich, whereby 'practical formulae' are indeed emerging from folklore and "beliefs of great antiquity."

---

235 Margaret Baker, *Gardener's Magic And Folklore* 1978 p.42.

6. Star Rhythm in Crop Yield

One star differs from another star in its glory...
- I Corinthians, XV, 41

This chapter presents a full account of British sowing trials where this author was involved. It groups them by the four elements, for ease of perusal, although it may lose the chronological sequence somewhat. Just two experimenters, Reg Muntz and Colin Bishop, performed the trials mainly in the years 1975-8, with some help from the writer. The yield effects were in what Biodynamic gardeners call root-days (Earth), fruit/seed days (Fire) and leaf-days (Water).[1]

The Muntz experiments all used a similar planting schedule, where one row of vegetables was sown per Moon-zodiac division, i.e. twelve rows per 27 days; Bishop's first experiment (1976) was of this form, but after that he sowed daily or twice daily since he was doubtful about the calendar involved. His data is given in an Appendix. Yields were computed as total weight per row, but on occasion the number of crops per row varied erratically, due for instance to insect attack, when the yields were computed by dividing the yield per row by the number of crops grown, to give mean weight per plant for each row. A successful experiment should show the star-rhythm using either method: per row or per plant.

Earth - and Potato/Radish Yields

In 1976 a test of Thun's theory was made by Reg Muntz, a professional market gardener in Sussex.[2] This gave interesting results despite the fact that 1976, which was the driest summer for 300 years, may not have been the best time for the testing of Moon influences upon vegetable growth.

One row of potatoes was sown each time the Moon passed through a constellation, over a period of two sidereal months from April 7th to May 28th. So in all 24 rows were sown. Each row was ten feet long and sown with ten seed potatoes. The weight of the ten seed potatoes was recorded before sowing. Harvesting was done on one day in September, by which time the potatoes had finished growing. The total weight yield for each row

---

[1] 'Lunar Calendar for Farmers and Gardeners', *Mercury Star Journal,* Midsummer 1977. The raw experimental data for this chapter may be found on the Considera website, go into 'Planting by the Stars' and then 'Worked examples' and 'Entering data' – for the latter you may need to log in. For a review of evidence for lunar calendars, see Martin Gardiner article.
[2] N.K., *Mercury Star Journal,* 'Zodiac Rhythms in Plant Growth (Potatoes)' Easter 1977, reprinted in Biodynamics, US, Winter 1993 'Testing the Lunar Calendar' pp.44-48.

was recorded. The results are summarized in the table, using the yield ratio: Weight of potatoes harvested per row / weight of seed potatoes sown per row.

Table 1: Yield Ratios for Muntz Potato Experiment 1976, grouped by Elements

Air	Water	Fire	Earth
3.0	4.8	4.7	4.7
2.5	4.5	6.7	9.1
5.1	4.4	7.2	7.5
7.9	4.7	6.0	6.4
3.7	3.5	2.4	3.9
1.7	2.1	2.6	2.2
	mean	values	
4.0	4.0	4.9	5.6

They have been grouped according to whether at the time of sowing the Moon stood in an Air sign, a Water sign, a Fire sign or an Earth sign/constellation. This shows that each row gave around five times the weight of the seed potatoes sown. To compare the Earth or 'root-day' sowings with the rest, the data in Table 1 may be summarized as follows:

Root-day yields / Others
5.6 ± 2.3 (n=6) / 4.3 ± 1.8 (n=18)
- a 30% excess, where 'n' is number of rows sown

Thus, 2.3 is the 'standard deviation' of yield ratios from the six 'Earth'-day sowings The mean value for the six root-day sowings was thirty percent more than for the remaining group of eighteen other sowings. For the rest of this chapter the results of experiment will be summarized in this manner; the data as in Table (1) may be found on the 'considera' website.

These results are expressed graphically in Fig. 1. The variation in yield is partly a seasonal change, due to the Sun's influence. For a statistical test, the results have to be corrected for this seasonal change. This was done by calculating what is known as a moving average, shown in fig.1. It is a 'five-point' moving average, where each point is a mean of five successive yield ratios.[3] When the moving average was subtracted from the yield ratio for each sowing the difference appeared as either a positive or negative deviation from the moving average (fig.2). These deviations were then grouped according to zodiac constellation and then the average for each sign taken (fig. 3).

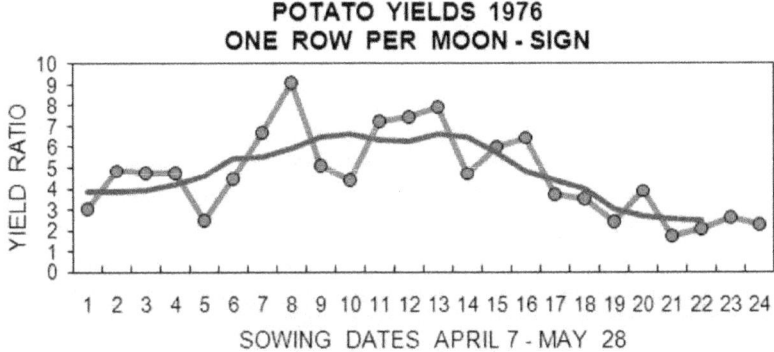

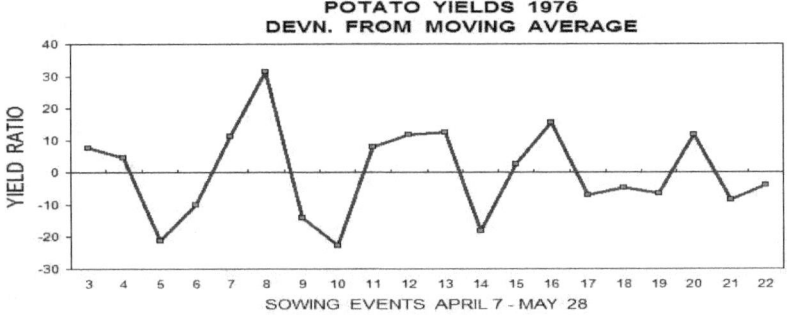

Fig 6.1, 6.2, 6.3: Yields from 24 rows of potatoes sown by Reg Muntz in 1976, given as weight yield per row / weight sown per row plotted by sowing date. After subtracting out a five-point moving average, the values thereby obtained are grouped by the twelve sidereal Moon-signs at sowing data and averaged. Maxima appear in the three Earth signs.

The significance test here used compared two groups: that of the element for which an excess has been predicted, and a second group containing all the others. The deviations from the moving average gave a mean deviation

---

[3] The Mather lunar-gardening trial of 1940 (Appendix 1) was the first to use a five-point moving average for the seasonal trend.

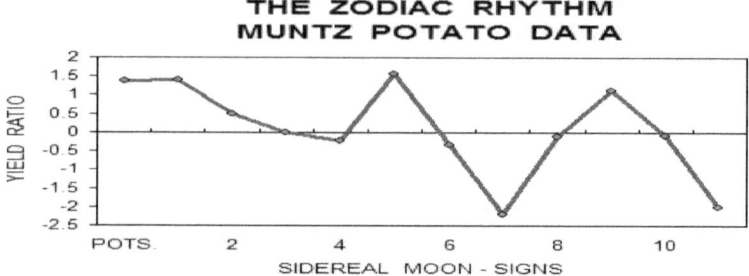

for the five earth signs sowings of 1.6 + 0.9 and a mean deviation for the fifteen sowings in the air, water, and fire signs of 0.5 + 1.1.

Statistically, the difference between these two means is very significant. The probability of getting such a difference by chance is only about 1 in 500. We find that out by applying a t-test for significance gives, which gives t=2.7, then tables will then convert that into a significance level).

These findings indicate a significant increase in yield for potatoes sown in Earth signs. The results suggest the presence of a periodic variation, with a maximum in each earth sign and a minimum in each water sign. This wave pattern, a 'third harmonic' of the sidereal lunar month, is comparable with the effect described by Maria Thun.

The Bishop Radish Trial of 1978

Mr Colin Bishop is a civil engineer and a member of the British Astrological Association, whose interests include gardening and the mathematical use of harmonics for investigating astrological influence. He began these experiments using lettuce, the results of which will be described in the next section. Then in 1978 he sowed radish seed twice a day for over a month in his allotment in Cardiff. Four different rows were sown each day, two in the morning and two in the afternoon, each time sowing fifteen radish seeds in five foot rows. In all, 156 rows were sown over 39 consecutive days.[4,5] Each row had comparable soil and light exposure. The radish were all harvested after only six weeks of growth, and received no watering, so that the yields were small.

I view this as the most significant Moon-gardening experiment yet conducted. Not one morning or evening sowing was omitted, nor were any rows eaten by birds or slugs. Such continuity is vital for statistical analysis. Mr Bishop's decision to sow one hundred and fifty-six rows resulted from

---

4 N.K. 'A Lunar Sidereal Rhythm in Crop Yield and its Phasing in the Zodiacal Circle' Correlation June 1981.
5 N.K. 'Star-Rhythm in Crop Yield,' *The Astrological journal,* Spring 1985, p.118.

questions which arose from earlier experiments concerning the nature of the celestial influence operating in vegetable growth. Thanks to this experiment, vital questions can be answered that had previously remained conjectural.

In the following data analysis, the two a.m. sowings were averaged together, as were the two p.m. sowings, giving in all 78 rows. The Figure shows the yields, for a.m. and p.m. sowings, with the trend line, and also Sidereal Earth Moon-signs indicted. Lunar phase does not have any effect is linked with it. If we group the data as Earth-day sowings versus the rest, this gives:

> Sidereal - Earth / Others
> 3.14±1.9 (n=18) / 2.11±1.5 (n=60) oz - a 49% excess

That is a large effect and one which is highly statistically significant. We derive the significance of this result by the same method as was earlier applied on the Thun-Heinze potato data, comparing the root-day sowings

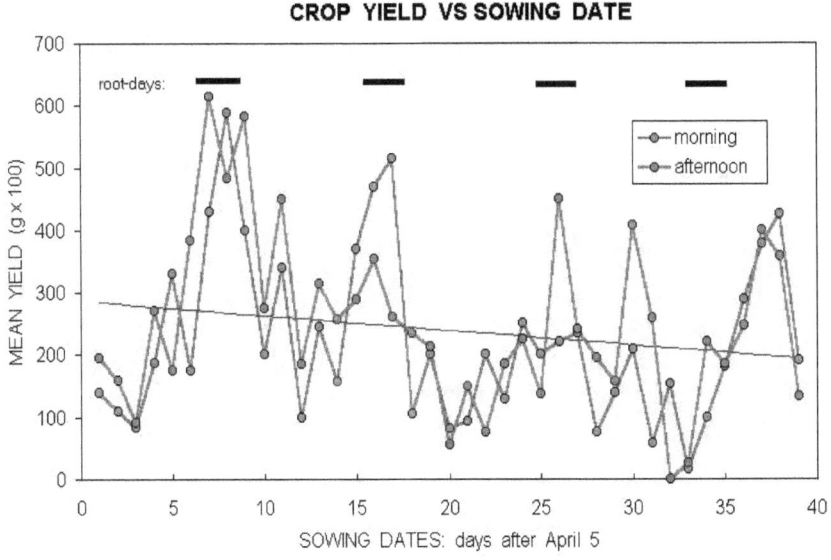

Fig 6.4: Radish by Colin Bishop, 1978   Yields from morning (dotted line) versus evening sowings over thirty-nine days, each point being a mean of two yield-values.

with all the others. This gives a significance of one in a thousand, i.e. there is a one in a thousand likelihood of getting such a result by chance. The 36 rows sown in root-days had a mean yield of 3.14±1.9, while the remaining 120 rows averaged 2.11±1.5. This gives a t-value of 3.8, and using tables this t-value gives a chance likelihood of 1 in 1000. If, alternatively, we took the opposite element group of Water, i.e. leaf-day sowings, and compared that with the root-day excess, a rather higher level of significance would be obtained.

If instead we divide the data using the unequal BD constellations, this gives:

BD Earth / Others

2.84±1.9 (n=23) / 2.15±1.5 (n=55) oz - a 32% excess.

The number of Earth sowings has greatly increased, though the excess in this element has dropped considerably. An alternative constellation division has been boldly devised by Mr Paul Platt in America, and used briefly by the Kimberton Hills calendar, discussed later on. This gave :

Platt Earth / Others

3.02±1.9 / 2.12±1.5 - a 42% excess

which is an improvement on the result using the BD divisions. His divisions are less irregular than the BD constellations.[6]

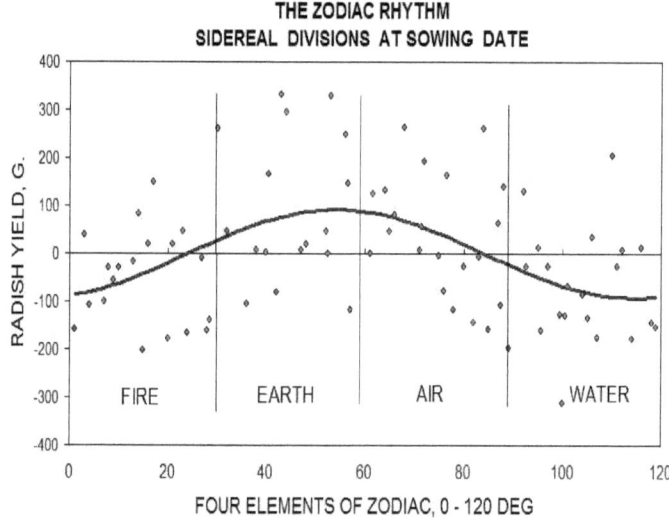

---

[6] To perform the several 'four element' divisions, zodiac longitudes of lunar position were determined for all of the radish-sowings. Then, longitudes of the different divisions as given in the Kimberton Hillls Calendar were used to allocate the four-element divisions. For example, in the tropical (astrological) reference, the sign of Aries is 0°-30°. The zodiac

<u>Fig 6.5</u> The Bishop 1978 radish data plotted by 120° of celestial lunar longitude, to show the Four Elements, with best-fit sinewave.

Or, we can choose not to use any constellation divisions, and figure 5 shows this approach. We take the Moon's celestial longitude at time of sowing, then subtract 24° from the given 'tropical' longitude (in the ephemeris), and that gives 'sidereal' longitude (see next chapter), then we reduce those longitudes to a 120° 'third harmonic' dial, so that, for instance, 130° would become 10°.

A best-fit waveform is then put through it. That is a sidereal waveform, a star-rhythm, and it's a third harmonic of the sidereal lunar month. Its amplitude gives us a different way of estimating how big the effect is, instead of the tabular divisions we have just looked at. Its amplitude is 36% of the mean yield. We can see how the peak of this best-fit waveform isn't quite centered on the 'Earth' trigon, as it 'ought' to be.

This Bishop data enables us to test the tropical zodiac 'water-sign' hypothesis, as used extensively in Moon-gardening columns and almanacs. Experiments which involve sowing in the middle of a constellation cannot do this, because they assume a specific zodiac system in their design; nor, on the other hand, can the experiments described in Rhodden's 'Planetary Planting,' due to reasons described earlier: merely two sowings are inadequate. Comparing tropical Water with all other sowings gave:

<u>Tropical Water / Others</u>
2.58±1.3 (n=21) / 2.32±1.4 (n=57) an 11% excess

There is a slight excess here, however the main excess fell in Air signs, with a minimum in Fire signs. No-one has ever predicted maxima and minima of this form. Symbolically, it makes no sense.

It is clear that the Thun hypothesis was supported by the Bishop radish data, and that the equal-interval Sidereal reference worked a lot better than the BD divisions. In addition, this data showed a greater yield for pm sowings (13%), as compared with am sowings, supporting traditional gardening lore. The next chapter returns to this set of data.

Fire - and Bean Yields

Three years of broad bean trials were conducted over the years 1975-1977.[7] The author collaborated with Reg Muntz, in a series of experiments

---

position of Aries in a Sidereal reference would be 24.5–54.5°, because of the 24.5 phase-difference between the two systems. The Bio-dynamic divisions use 29°- 53°, allowing the Ram constellation a mere 24°, while Mr Platt's divisions (printed in the 1989 Kimberton Calendar), are 22°- 52° ie 30° for Aries.
7 N.K. 'Zodiac Rhythms in Plant Growth, III: Beans' Star and Furrow, Winter 1984.

to test the Thun theory. They were performed at Mr. Muntz's 'Barnards Nursery' near Pulborough, Sussex. The experiments differed from those reported by Thun in being performed on poor quality soil, which had only had grass growing on it previously. They are of interest as regards the question, to what extent would the 'Thun effect' manifest on crops grown in poor soil? This aspect of the experiments came about through necessity rather than choice: a market gardener needs his good soil for produce, and the experiments cannot easily be designed in such a way that what is grown over a number of rows is available for consumption when required.

In these experiments, summarized at the end of this chapter, twenty-four or thirty-six rows were sown of a particular crop, sowing one row for each Moon zodiac sign, approximately every two days, i.e. each day when the Moon was near to the middle of a zodiac constellation. Twelve rows are a minimum number that can be sown for a Moon zodiac experiment, but multiples of this were taken in the hope that longer, monthly Moon-rhythms might also show up. A sowing timetable was drawn up for the sowing months. In such experiments, crops can be either harvested after they have all grown for the same period of time, over a two or three month program, or all on one day when they have all finished growing. In the case of broad beans, the latter course is less trouble, but means leaving them to dry out so they cannot be eaten.

Seeds were sown sufficiently thinly that later thinning out of crops was unnecessary, as it is rather difficult to thin out to the same degree for every row. Likewise watering of the rows was kept to a minimum and avoided where possible. Fresh weight yields of pods per row and the number of plants grown per row were tabulated as results. Sometimes the whole plant was weighed, as total weight per row, to ascertain whether the weight ratio of pods to rest of plant varied with the element of the zodiac.

Between March 23rd and May 15th 1975, twenty-four rows of broad beans were sown. The rows were each five foot long and ten broad beans were sown in each. A plot of land ten by twenty foot was used with comparable soil and sunlight exposure across it. Most of the seeds grew, giving between six and ten bean plants per row. The pods were all harvested and weighed on a day in August. Some were kept for sowing next year, stored in "zodiac packets" so that they could be sown in the same zodiac sign as the plant from which they had been harvested.

A similar experiment was performed in 1976, but with 36 rows of beans instead of 24. Beans from the previous year's experiment were used as seeds. In 1977, 24 rows were grown. In both years the seeds were partially eaten by birds, which ruled out taking the total yield of beans per row. Instead, the row yields were divided by the number of plants which grew in each row, giving mean crop yield per plant for sowing date. 1976 was a year of severe droughts, giving wide scatter to the data. An alternative in the case of a bean crop experiment, as used by Thun, is to pick beans from each row

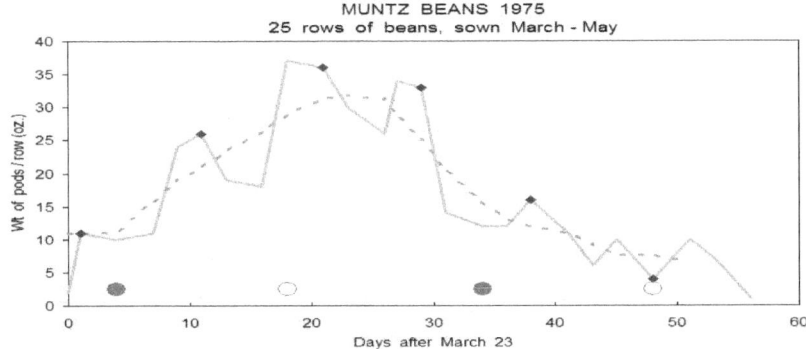

Figure 6.6: The Muntz Bean Experiment, 1975 The figure shows a large seasonal effect in the yield data. The weight-ratio of pods compared with the rest of the plants (taking "rest of plant" weight after the beans had been removed, the plants pulled up, and the roots washed, per row) showed "fruit day" sowings with a slightly higher weight fraction for beans than the other three elements or "trigons."

when they are ready, recording weight per row each time; so that final weight yields per row are the sum of several pickings. The Seville Longpod variety was used, from Chase Compost Seeds Ltd. Suffolk.

Taking total fresh weight of pods per row, in ounces, over the 24 rows, and grouping by Fire Moon-signs, i.e. fruit or seed days versus all the others, gave for 1975:

Fruit/seed day sowings / Others
Weight yields: 21.0±11 / 16.3±9.7 oz - a 29% excess
Wt. ratio pod/rest of plant: 40 ± 9% / 34 ±8 % - a 17% excess

For the 1976 bean experiment, sowings commenced on Jan 22nd and finished on April 11th, sowing one row per zodiac sign:

Fruit/Seed day sowings / Others
9.93 ± 4.2 oz (n=9) / 6.8 ± 3.2 (n=27) a 46% excess

The yields from the 1977 broad bean experiment were from rows sown over two sidereal months, as the mean yield of pods in oz. per plant. Sowing commenced on Jan 2nd and finished on Feb 24th:

Fruit/Seed day sowings / Others
2.46 ± 0.26 (n=6) / 1.89 ± 0.59 oz (n=18) a 30% excess

The yields for beans sown on 'fruit-days' (i.e. when the Moon was traversing one of the three fire-signs of the zodiac) were significantly higher than yields on other days, overall. To see how significant the results were, a test was applied to the data. This test compares differences between the means of two groups, one being in this case the yields from fire-element Moon-signs, and the other group being all the other yields. The difference between these two groups is assessed for significance in terms of the likelihood of such a result being generated by chance.

The 1975 experiment had a large seasonal variation effect running through it which first had to be removed by subtracting out the moving average from the data. Of the remaining 22 data points, five were fire-element sowings, and their mean was significantly above the other at the 5 per cent level, i.e. there was a 1 in 20 likelihood of getting such a result by chance. The 1976 and 1977 results combined were significant at nearly 1 in 1000.

For three successive years, broad beans sown in relatively poor soil gave significantly increased yields for rows sown when the Moon was in a fire sign at their sowing. This accords with Thun's theory which predicts that the last stage of a plant's growth, when the seed is formed, has an affinity with this trigon. The average increases in the weight yield of beans for the fire-sign sowings, compared with those from the other sowings, were:

29% in 1975 (24 rows sown)
44% in 1976 (36 rows sown)
30% in 1977 (24 rows sown)

There was a rather large random scatter in the results, especially in '76 and '77. It is not clear whether the 'zodiac breeding' over three years had any effect.

Water - and Lettuce Yield

In 1976 Mr Bishop sowed 36 rows of lettuce seed using the Rulni calendar to tell when the Moon was near the midpoint of each constellation, planting in succession one row per sign over three sidereal months.[8,9] The soil used for this experiment was a quickly draining silt, fairly low in humus content. Lettuce was used chiefly because Culpeper's 'Complete Herbal' says of lettuce that "The Moon owns it". The lettuce were harvested in rotation, using a similar schedule to that for sowing, so that each row grew for a similar period of time. The weight of lettuce was taken after cutting off the roots.

---

8 Colin Bishop, 'Moon influence in lettuce growth,' *The Astrological Journal,* Winter 1977.
9 N.K. 'Zodiac Rhythms in Plant Growth: Lettuce,' *Mercury Star Journal,* Spring 1978 (compared zodiacal & constellation divisions for the Bishop 1977 and 1978 experiments).

The Thun theory predicts that the Water Moon-signs should have an excess in lettuce yield. Taking the nine rows sown at these times, and comparing their final yields with all of the others in the experiment, gave:

Leaf-day sowings / Others

131± 63 / 87 ± 54 gms - a 51% excess

After subtracting out the moving average the two groups became:

Leaf-day sowings / Others:

39 ± 37 (n=8) / -12 ± 31 (n=24) gms

which is significant at 1 in 50,000. Just as significant a result could be obtained from the yield decrease in the group of root-day sowings (mean value -36g ±31). A statistically more significant result would be obtained by comparing these Water and Earth groups in a t-test.

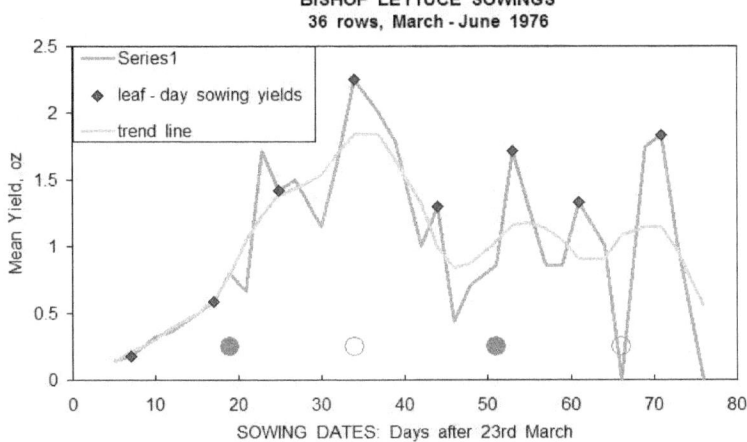

Fig 6.7: Bishop Lettuce 1976 - The Figure shows the mean yields for lettuce tops per row after a two month growing period. The graph well shows the sidereal lunar rhythm weaving around a seasonal effect. A five-point moving average was put through the data, representing the seasonal trend. Subtracting this out has the effect of losing the first two and last two data points, giving 32 in all.

Use of the BD constellation divisions would give the same totals, since sowings were made with the Moon near the middle of each constellation. These results could equally well be described by saying that minimum yields are obtained with the Moon in sidereal earth-signs. There appear to be equal and opposite effects, alternating maxima and minima, in short a wave-pattern. Tropical zodiac totals show the sidereal means simply displaced one sign without appreciable change, so that maximum yields now appear in tropical fire-signs instead of sidereal water-signs.

These 1976 results suffered from the severe drought conditions that year, which caused the yields to be small. The next year's results were more substantial owing to better weather and also because this time each row of lettuce was allowed a three-month growing period instead of two.

Mr Bishop was interested in the question as to which zodiac he was here dealing with, and therefore in his 1977 experiment instead of sowing one row per Moon-constellation he decided to sow them more frequently, one every two days or so. His efforts were not without success and deserve scrutiny. As one sympathetic to astrology and unfamiliar with B.D. farming, Mr. Bishop was naturally predisposed to believe that it was the Tropical zodiac that was here operating. From his 1977 experiment he hoped to be able to tell whether the Moon's position in Tropical or Sidereal zodiac gave a better interpretation of the results. This predisposition may be important, if, as we have seen, plants are affected by the attitude of the gardener. For the 1977 experiment a Rulni calendar was not used and sowings were simply made whenever convenient, noting the date and time of each sowing.

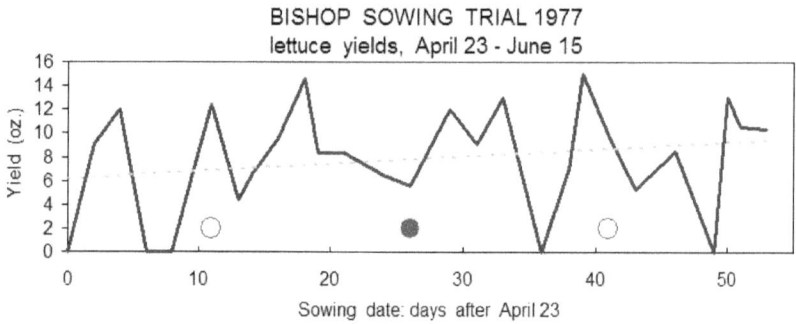

Fig 6.8   Bishop lettuce 1977, yield versus sowing dates after April 23

Early sowings failed to germinate owing to cold weather, and germination of the lettuce seeds did not begin at all until the latter half of April. Between April 25th and June 15th twenty-six rows were sown, most of which grew. Each row was thinned out similarly, received similar amounts of daylight, grew for the same length of time, and was not watered. The mean yields of the these rows, i.e. total weights of the lettuce tops after removal of their roots per row divided by the number of lettuce grown per row, were averaged as before over the Moon sign element.

Table 2: Moon-sign element means for 3 systems
Bishop lettuce 1977 yields in oz.

Star-Rhythm in Crop Yield

	Tropical	Sidereal	Biodynamic
Earth	8.5	5.1	5.1
Air	5.1	9.9	10.2
Water	10.2	9.5	8.1
Fire	8.1	8.0	8.5

By an unlucky chance, of 26 rows sown, only four fell properly inside the sidereal water-signs. Another two fell on its cusp[10] making six in all. Here the tropical (astrological) system has scored best! However, the tropical zodiac divisions give a different pattern from last year, with maxima in the water-sign sowings, not as last year in fire-signs. Sidereal Water has a net excess for the 1977 lettuce which is not statistically significant:

Leaf-Day Sowings / Others
9.5 ±5 (n=6) / 7.7± 4 (n=20) - a 22% excess.

Were we to score just the four rows falling well within Sidereal Water, and not those on its boundary, its mean yield value would be 11.0 oz., which is more clearly an excess. The results obtained using the constellation-zodiac (Rulni) fail to show any increase in water-sign sowings.

If the reader will allow a slight modification of procedure, let us turn our attention to the difference in yields between the two opposed elements, which in this case are Water and Earth, for leaf- and root-day sowings. The influence which makes a maximum effect on one of these elements causes a minimum four and a half days later, when an opposite tendency is working.

In other words, let us suppose that the decrease in root-days is as significant as the increase in leaf days for lettuce yields. The experimental data here presented often seems to confirm such a conjecture. We will return to it in the next chapter where a sine-wave analysis of the data is performed. Philosophically this idea is important for our appreciation of how the celestial influence works. In that case, what we may call the amplitude of the effect is given by the excess of leaf-day yields together with the deficit

---

10 'On the cusp': one row was sown at 7pm on May 4th when the Moon stood at 24.5 degrees Scorpio, which was inside the BD constellation of the Scales, extending from 9° to 27° of Scorpio (tropical). However, Sidereal Libra ends at 24.5° Scorpio, so this sowing just comes within the next sidereal sign, the water-sign of Scorpio. The convention here used was that sowings at 24° were included in the next sign.

of root-day yields, averaged together. For example, the mean Sidereal leaf-day yields from the above table averaged 9.5, while for the root-days it was a mere 5.1, and the mean value for the whole experiment was 8.1 oz, which gives a mean amplitude of 27%.

If this procedure is acceptable, it gives a value less dependent on the small number of sowings which happened to fall into Sidereal Water in Mr Bishop's sowing schedule.

The equal-interval sidereal zodiac gave fairly consistent results over the two years of the experiment. As with our earlier analysis of Bishop's radish data, the divisions advocated by Mr Paul Platt have been included. The above table includes, I believe, all the zodiac reference frameworks ever to have been used in a Moon-gardening context. We see the same general effect as with the radish data, with Platt's constellation boundaries scoring better than the BD ones. From this we would conclude that the best Kimberton Hills calendar yet produced was that of 1989, which used Platt's divisions.

Amplitude of Star-Rhythm for Bishop 1977 Lettuce

Framework	Water	Earth	% Excess
Sidereal	9.5	5.1	27%
Platt	9.7	6.1	22%
Biodynamic	8.1	5.1	17%
Tropical	10.2	8.5	10%

Mean yield: 8.1 oz.

Normally, a peak in sidereal Water would give a tropical peak in Fire, but it just happened that this data's distribution also gave one in Tropical Water. In addition, that year the lunar nodes seemed to be inhibiting growth very strongly. The figure shows that, out of the five sowings made near the lunar nodes four wholly failed to germinate, in accord with Thun's findings.

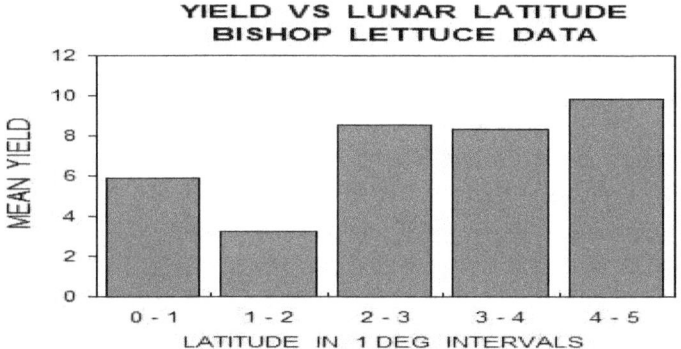

Fig 6.9   Bishop 1977 lettuce, plotted by lunar latitude.

1986 Lettuce trial

The following experiment has not been published hitherto, as it was on rather a small scale and failed to reach statistical significance. In 1986 Colin Bishop performed a lettuce experiment, first sowing indoors, then transplanting them into the soil when ready. As in previous years, he sowed one batch per day, continuing over 36 days, from March 19 to April 23. Each evening he sowed nine seeds into three pots, and six weeks later the best from each pot was planted out, giving usually three lettuce growing from each sowing day. As his account of the experiment explained, 'The intention was to make one sowing late each evening regardless, and without knowledge of the Moon's position.'

Thus, one reason for making sowings daily is that no attention needs to be given to what is happening in the heavens in case one's expectations may be viewed as influencing the result. Colin Bishop wrote to me:

> It is considered advisable to use only tools such as tweezers or small knives to handle the seeds so as not to touch them by hand, thus avoiding the "green fingers" effect. There is some evidence that hormones are transferred from the skin of some people to plants, and this is why some people are much more successful horticulturalists than others. Immediately after sowing, the seeds were given a thorough soaking and the time recorded, and this has been taken to be the 'sowing time'.

A problem in previous years' attempts of Moon-experiments upon lettuce growth lay in the different growing period through the season. If the lettuce were all picked at the same time, then the later ones would not have grown properly while those sown earlier would have tended to bolt, spoiling the experiment.

To overcome this, when harvesting of lettuce began, about ten weeks after sowing, the picking was done in sequence two rows at a time each day, so that the 34 rows grown were picked over a mere 17 days. This gave the last rows a 2-3 week shorter growing period than those at the beginning, giving more even yields through the experiment.

Seed germination was around 60%. Sowings were normally done at 10pm. On two days Mr Bishop was not able to sow any seeds, March 26th and April 18th, and on each occasion an extra set was sown at 6am the next morning. The weight of each lettuce head was recorded.

The 36 rows were sorted by the four sidereal Moon-sign elements as previously, and the element-means of lettuce weights obtained by dividing total weight of yield in each of the four element groups by the number which grew in each of these. That seemed the simplest approach. This gave:

Element: <u>Fire</u> - <u>Earth</u> - <u>Air</u> - <u>Water</u>
Mean yields: 122g - 107g - 136g - 146g
- For 9, 9, 10 and 9 rows grown -

If the Biodynamic constellations are used instead, we obtain:

Element: <u>Fire</u> - <u>Earth</u> - <u>Air</u> - <u>Water</u>
Mean Yields: 125g - 123g - 156g - 111g
- For 11, 12, 7 and 7 rows grown -

It will be noted that there are now more root-day (Earth) sowings than others, because the constellation divisions give more space to the former. The mean Water-element yield per row (leaf-days) is now less than the other yields! The unequal constellation-analysis has lost the predicted effect. There is a suggestion in this data of a minimum yield in the Earth or root-day sowings. As before, let us estimate the amplitude[11] of the effect using different zodiac divisions. Mr Bishop computed the values here used.

The experiment is the fourth occasion on which Mr Bishop has obtained a positive result, i.e. a yield increase in the element predicted to be in excess by the Thun model. It appears perplexing that an experiment with just over seventy lettuce, which did not reach statistical significance, can yet give some testimony concerning the divisions of the firmament. It shows what can be achieved with a good experimental design.

---

11 The amplitude is expressed as a percentage: (Water mean – Earth mean) x 50/total mean

Amplitude of Effect, Water/Earth

	Water	Earth	% Excess
Sidereal	146	101	18%
Biodynamic	112	123	-4%
Tropical	149	121	11%

Overall mean yield, 128 oz.

Five Hundred Rows

The Table below summarizes all the sidereal Moon gardening experiments in which I have participated, where the experiment has been properly conducted and completed. It specifies the person doing the experiment, Reg Muntz or Colin Bishop, then the year, the vegetable used, the number of rows sown, the overall percentage excess (the mean weight yield per row, in the predicted element, as compared with those in the other three elements) and the element for which that excess was measured. An asterisk shows those results which have been previously published. Up to four lettuce grew per row; taking mean yields per row, for each of the 36 rows, of which there happened to be nine per element, then averaging these row-means, gave:

Water / Fire / Earth /Air
138g / 107g / 109g / 122g

Five of the Muntz experiments have not hitherto been written up, as they were not statistically significant and on one occasion negative in result, but they are here included for completeness. These are numbered 2, 3, 6, 7 and 9.

The 1975 lettuce experiment had 26 rows sown in 5 feet long rows, which were thinned out to one lettuce per 9 inch, and each row was grown for 3 months, which was too long and they all bolted. The water-sign rows averaged 53% more by weight. If yields per row were divided by the number of lettuce which grew per row, giving mean weight of lettuce per row, then the figure became 36% more growth in water-signs. Despite this large excess, a wide scatter in the data meant that the experiment was not statistically significant. An experiment with 24 rows of potatoes was attempted by Muntz in 1977, but the soil was so poor that they hardly grew.

This sequence of experiments, the most comprehensive replication of the Thun-Heinze experiments yet published, shows consistently positive results. No especially good soil was used: the organic market gardener Mr

Muntz could not spare his best soil for such a purpose. I believe that no sowing trials have been reported using flower-type crops. These would presumably include such things as brussel sprouts, cauliflowers and broccoli.

The Muntz experiments

	Yr	Veg.	Rows	Xess	El.	Publish
1	'75	Beans	24	29%	Fire	*
2		Pots	24	3%	earth	
3		Lettuce	26	53%	water	
4	'76	Beans	36	44%	fire	*
5		Pots	24	30%	earth	
6		Carrots	32	13%	earth	
7		Lettuce	24	7%	water	
8	'77	Beans	24	30%	fire	*
9		Lettuce	28	2%	water	

The Bishop experiments:

1	'76	Lettuce	36	51%	water	*
2	'77	Lettuce	26	22%	water	*
3	'78	Radish	156	49%	earth	*
4	'86	Lettuce	36	21%	water	

Over the five years of these experiments, a total of 496 rows were sown. On average they showed a one-third increase in the predicted 'trigons' compared with others: a 49% increase for the Bishop data and 23% for the Muntz data. Alternatively the results from these five years of experiments can be expressed by saying that, if the phenomenon here quantified were of a general nature, then a person using this four-element star-rhythm should expect to obtain 22% greater yields than one not using it.

7. The Starry Script

On either side of the river grew the Tree of Life, bearing Twelve kinds of fruit, a different one for each month.
- Book of Revelations: 22,2.

Every two or three days, the Moon enters a new constellation. As long as farmers sow their crops near to the centre of these periods, then disputes over constellation boundaries can be ignored. One hears this sensible and pragmatic view expressed by Biodynamic gardeners. Biodynamic farmers use a phenomenological approach, observing (at least in theory) the visible constellations past which the Moon moves in the night sky. In the 1940s when the 'Rulni calendar,' the precursor to the modern Thun calendar, was composed, few had heard of the Sidereal zodiac, with the ancient tablets still in the process of being deciphered. A diagram of the star-zodiac is here shown, with its 30° divisions as used in antiquity.

If the Thun four-element theory is accepted, then the experimental data demonstrates that a sidereal reference of some kind is operative. What choices are available? This chapter reviews those sidereal frameworks, which have been used horticulturally.

A Confusion of Terms

Publications concerning lunar gardening normally present a choice between two celestial reference systems. Here, for example, is the choice as given by the Lllewellyn almanac:

> A key difference between the two systems is that the tropical zodiac is divided so that each sign is exactly 30 degrees, while the sidereal is divided unevenly, according to the size of the stellar constellation.[1]

The former being the recommended framework. In publications such as John Soper's 'Biodynamic Gardening' or 'Biodynamics' from N. Zealand, the same duality is presented, between unequal constellations or equal tropical-zodiac signs, though the recommendation as regards which to use is the reverse of that given by Llewellyn.

Of the two options, one uses signs and the other uses constellations, one is tropical while the other is sidereal, and one is astrological while the other is astronomical. Neither of these frameworks are here advocated for deciding when a crop should be planted. We have seen how crop yield increases generally show up better in the element predicted by the Thun theory, if

---

[1] Llewellyn's *Lunar Organic Gardener*, 1993, p.89.

analyzed by the equal-interval Sidereal zodiac of antiquity, rather than by the unequal constellations. It may be helpful here to use an upper-case S for Sidereal as referring to the zodiac system.

Hitherto we have used the terminology of 'Moon-constellations,' but really we should have referred to 'Moon-signs,' taking these as being sidereal. Let us move away from the illusory Biodynamic (anthroposophical) notion that an influence is transmitted via the unequal constellations. Historically, no cosmologist has ever believed such a thing, or at least not until quite recently. Let us not confuse the label with that which is labelled. The constellation-pictures in the sky appear as of unequal lengths, but this does not imply that the patterns of celestial influence operate in twelve unbalanced and unequal sectors. Traditionally, the term 'zodiac' applied to an equal twelvefold division of the ecliptic. In this sense, Biodynamic farmers are not using a zodiac, but are instead using the constellations. Historically, the four elements were applied to the signs, ie the equal twelvefold divisions of the zodiac, never to the constellations. It is only BD farmers who have applied the qualities of the four elements to the unequal constellations.

Because BD farmers always apply the elements to the unequal constellations, we have used this framework in explaining the Thun model. As the term 'sign' is nearly always used to refer exclusively to the Tropical system (at least, in the West), it has hitherto been avoided. I didn't like using such a terminology, as there is no warrant from tradition or anywhere else for applying the elements to the unequal constellations. It would be greatly preferable to use the term 'sign,' e.g. a Water-sign, pertaining to a sidereal reference, in this context.

As the term 'sign' is ordinarily used as pertaining to the tropical zodiac, one hopes this will not cause confusion. It refers to a system that uses 30° divisions. Here it will normally pertain to a sidereal reference because that is the one mostly discussed in the present work. This involves shifting the 'signs' from the usual astrological zodiac onto the star-zodiac, a rotation of about 25°.

A historical excursus now follows to show the 'glorious heritage' of siderealism, viz. the forgetting and rediscovery of man's original zodiac. Its theme is threefold, dealing with the zodiac constellations, the tropical and the sidereal zodiacs. It aims to show the metamorphoses which each of these has undergone, in the course of time.

The Forgotten Zodiac

<u>Fig 7.1</u> The Two Zodiacs and the Constellations

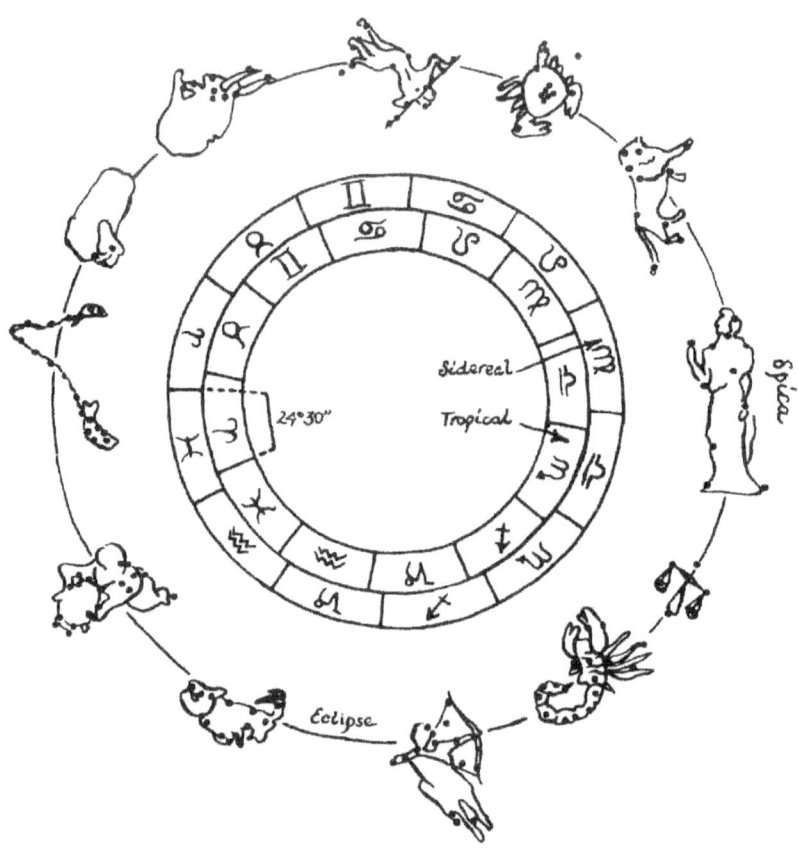

In the 5th century BC the Greek historian Herodotus visited Mesopotamia, where dwelt the Chaldeans. Herodotus was mainly impressed by their harvests. Where now the dusty desert blows, he witnessed harvests of such abundance as, he feared, would strain the credulity of his readers.[2] Surely, he gazed upon the fabled 'Hanging Gardens of Babylon.' The Babylonians had constructed an extensive and sophisticated irrigation system using canals:

> Of all the countries that we know there is none which is so fruitful in grain. It makes no pretensions indeed of growing the fig, the olive, the vine, or any other tree of the kind; but

---

2 The History of Herodotus, New York, 1928, p.72-3.

in grain it is so fruitful as to yield commonly two-hundred-fold, and when the production is the greatest, three-hundred-fold. The blade of the wheat-plant and barley-plant is often four fingers in breadth. As for the millet and the sesame, I shall not say to what height they grow, for I am not ignorant that what I have already written concerning the fruitfulness of Babylonia must seem incredible to those who have never visited the country. The only oil they use is made from the sesame-plant. Palm-trees grow in great numbers over the whole of the flat country, mostly of the kind which bears fruit, and this fruit supplies them with bread, wine and honey.

No-one told him that the zodiac was in the very process of being invented in that country. Did the astral science developed by the Chaldeans, for which they became far renowned, have relevance to the great fertility of their land?

Mesopotamia was the land of the two rivers, the Tigris and Euphrates. Between these the concept of the zodiac, the 'circle of animals' was born. A millennia earlier, the Sumerians in the same location had one version of the year containing 360 days, thirty days for each month of the year, and they developed a base-sixty arithmetic, which helped the framing of its divisions. In those pre-zodiac times they counted between fifteen to eighteen irregular-sized constellations lying on and around the ecliptic. A good half of today's constellations were there unchanged with the same names, such as the Lion and the Great Bull of Heaven, though others such as the Great Swallow and a Lady of the Heavens (both in the region that became Pisces) have vanished. They had a different figure for Aries (called The Hired Man or the Agricultural Worker), and a long figure corresponding to the Virgin (which seems to have been called 'The barley-stalk') occupying a good 40°, next to Libra the Balance nearby that was much shorter. The Moon moved round these dwelling-places of the gods as they were called, every month.[3],[4]

Towards the end of the sixth century BC, the Chaldeans developed a calendar of twelve months, named by the constellations. In the fifth century BC they divided the ecliptic into twelve equal thirty-degree divisions, creating the zodiac. Thereby it became possible to specify where the Moon

---

[3] Christopher Walker, 'A Sketch of the Development of Mesopotamian Astrology and Horoscopes,' Clio and Urania Confer, Ed Kitson,1989, pp.7-14, 9.
[4] Michael Baigent, *From the Omens of Babylon, Astrology and Ancient Mesopotamia* 1994, p.175.

and planets were in their motions. Positions are given in degrees, from zero to thirty, for the main zodiac stars.[5,6]

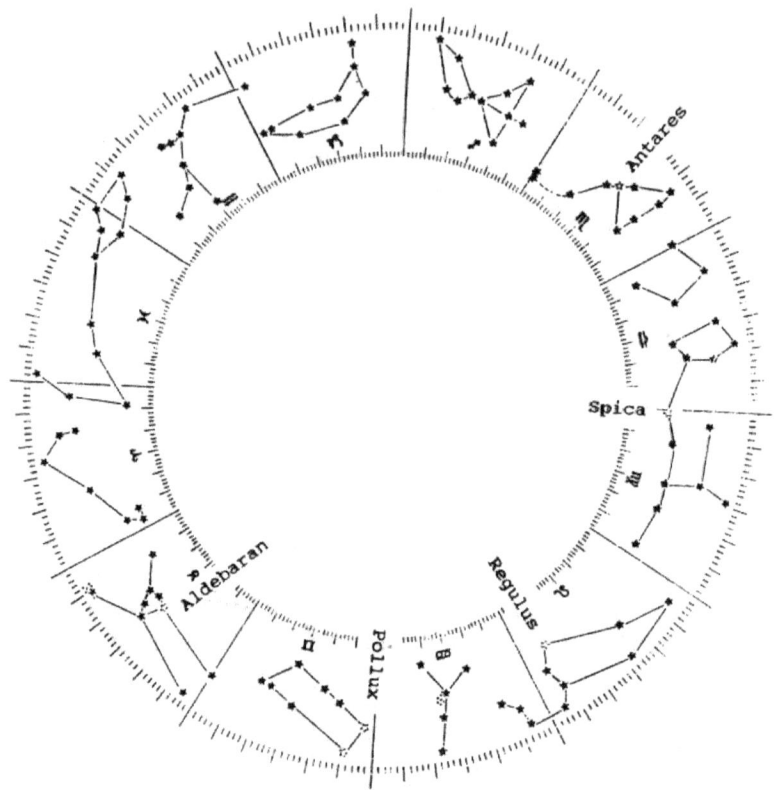

Fig 7.2 The Star-zodiac (after Powell) showing some fixed stars

In the Chaldean section of the British Museum, some clay tablets record this birth-process. One tablet, dated to the seventh century B.C., shows seventeen lunar 'mansions' around the ecliptic, and these were unequal constellation-images which had endured maybe for millennia. In the sixth century B.C. the twelve lunar months begin to impinge on this pattern, with

---

5 *The Zodiac: A Historical Survey* by Powell and Treadgold, (Anthroposophical Publications, Temple Lodge Press, republished by Astro Computing Services, California) 1985.
6 'For Babylonian and early Greek astronomy the beginning points of the signs are rigidly connected, not with the equinoxes, but with the fixed stars,' B.L. Van der Waerden, *Science Awakening* II, 1974, p.222; Marie Delclos, 'Astrologie racines secrètes et sacrées' Paris, 1994, Ch. 17; Peter Huber, in 'Ueber den Nullpunkt der Babylonischen Ekliptic' *Centaurus*, 1958, Vol. IX, 192-208 found that star-longitudes on ancient tablets agreed fairly well with that given by placing Aldebaran and Antares at the midpoints of Taurus and Scorpio.

each month having the name of a constellation beside it. A couple of the months had two constellations, so they had not yet reached the twelvefold division. Then, a century later (the fifth century B.C.) an amazing thing happens: tablets start to give the longitudes of stars, from zero to thirty degrees, in twelve signs. There are however no instructions as to where these twelve equal divisions are placed, we just have to gather that from the longitudes of the stars.

The bright star Spica signifies the sheaf of corn held by the Virgin, so that Virgo appears as some kind of harvest-goddess, holding the wheat. Spica is positioned at the end of the Virgin, say 28-30 degrees. Chaldean measurements of celestial longitude were within, say, one degree.[7] The star Spica seems generally quite important, but also and rather dramatically the red star Antares is directly opposite the pale-pink star Aldebaran in the heavens, exactly to within a couple of arc minutes in longitude. From where the Chaldean Magi observed on their ziggurats, did these two appear as special markers in the wheel of the sky? They are two of the brightest stars in the zodiac, and right on the ecliptic, and were two of the four 'royal' stars of ancient Persia, the other two being Spica and Formalhaut. These appear in the old cuneiform tablets as placed fairly near the centre of the signs of Taurus and the Scorpion.

The notion of the triplicities ('trigons') appear in early Chaldean texts. The constellations of Leo, Aries and Sagittarius were perceived as sharing some kind of energy in common, as did Cancer, Scorpio and Pisces. Centuries later, this was linked to the Greek four-element theory, one set being designated as fiery and the other watery.[8]

The zodiac of the Chaldeans shared in common with that of Ptolemy, that it did not quite contain the four elements. The four elements had been in use amongst the Greeks for many centuries, in philosophical and medical contexts, before they acquired a heavenly significance. They entered the frame of the zodiac at the end of the second century AD, at which time - let's be clear about this - the zodiac was still sidereal. The name of the Syrian Vettius Valens is associated with this elemental act.

The zodiac of the Chaldeans was adopted throughout the Hellenistic world of the Mediterranean in the early centuries AD, as the new art of astrology spread like an epidemic. It used a sidereal zodiac, not a tropical one: 'Hellenistic' horoscopes (ie, written in Greek) survive from the first to the

---

[7] N.K., 'The Star-Zodiac of Antiquity' *Culture and Cosmos*, Autumn 1997, pp.5-22.
[8] F. Rochberg-Halton, 'New Evidence for the History of Astrology,' Journal of Near-Eastern Studies, 43, 1984, pp.115-127, 122.

fifth centuries AD, composed around the Mediterranean, chiefly Alexandria,[9] and these were sidereal, i.e. anchored in position by the stars.

The science historians Otto Neugebauer and Van der Waerden have described this process, which is otherwise widely ignored and forgotten. The longitudes given in these charts are compatible with the star Spica being positioned at 30° of Virgo, as concurs with the present-day Indian version of the Sidereal zodiac. Spica was most important for the Egyptians, especially towards the end of the second century BC (as the 'Age of Pisces' began) when the equinoxes were pointing directly at it.

Twelve Constellations

Thus, the Chaldeans never had twelve unequal constellations. Early on, they did have unequal-sized constellations, but more than twelve, somewhere between fifteen and seventeen, wrapped around the circle of the ecliptic. A tradition of twelve unequal-length constellations first surfaced in the Greek-speaking world. The first star-maps of the twelve constellations are credited to Hipparchus of Rhodes, in the second century B.C. His maps were lost, but not before they were used by Ptolemy four centuries later, in Alexandria. Hipparchus measured his star-longitudes from the Vernal Equinox, i.e. he used the tropical zodiac. No-one took much notice of this procedure until Ptolemy adopted it. So the earliest ancestry of the twelve unequal constellations derives from the lost star-map of Hipparchus, expressing a folk-tradition passed down by sea-faring peoples and farmers who used the constellations.

Following folk-beliefs of his time, Ptolemy in the Almagest accepted that the constellation of the Virgin should be laid out horizontally, extending her over more than forty degrees of the zodiac. What became the adjacent constellation of Libra thus had to be rather short. In the Chaldean zodiac, standing decently upright as a corn goddess ought to do, she had occupied a mere 30° of celestial longitude. Ptolemy wrote an astrological book, the Tetrabiblos, which used twelve equal signs, and an astronomical work, The Almagest, which used constellations. The latter he perceived as being of unequal size. It never occurred to Ptolemy to specify boundaries of the constellations, or to use these unequal-sized constellations to describe celestial influence. Nor did it occur to any cosmologist since to conceive a

---

9 Otto Neugebauer & van Hoesen, *Greek Horoscopes*, Philadelphia 1959, p.594: '...the astrological literature of the hellenistic-Roman period still preserves the norm of Babylonian astronomy.' I found that seven charts in Neugebauer from AD 40-140 had a mean ayanamsa of 3.7±1.4 degrees, while for eleven charts AD 460-500 the equivalent was -2.0±0.4. This confirmed Neugebauer's view that the reference used was moving with respect to the tropical framework, ie that it was sidereal, and that they coincided momentarily in the second century.

circle of celestial influence as divided into irregular domains, or at least not until Biodynamic calendars arrived on the scene.

The Tropical Zodiac of Ptolemy

This work isn't concerned with the 'turning' or tropical zodiac - not at all. That is the one used nowadays by astrologers. Historically this zodiac started life as a Greek calendar-system somewhere in the 5th century BC, being linked to the four 'turning-points' of the year. The solstices and equinoxes are at zero Cancer/Capricorn and Aries/Libra, and always will be. This zodiac has no connection with the stars. Let's try to define it.

The Vernal Point is the Sun's position on the ecliptic (i.e. the line which the Sun's path traces in the sky) at the Spring Equinox. That position is more or less unobservable, and is quite an abstract concept, as one cannot see the Sun's position against the stars. The Chaldeans had little interest in its position against their zodiac. Ptolemy however fixed his zero Aries position to this point, making it the start of his zodiac. The divisions of the star-zodiac were then within a degree of this position. The rather awesome moment of synchrony finally arrived in the century after Ptolemy.[10]

The two zodiacs were conjunct somewhere around AD 250, one can't be more definite. At this point the fish became adopted as the symbol for the Christian religion, at the end of the second century, as the Vernal Point transitioned from Aries into Pisces. Spica, the star on the sheaf of wheat held by a fertility-goddess, as became the virgin, traditionally associated with fertility and prosperity, with good luck and good fortune, was then on the autumn equinox.

The signs have acquired agricultural meanings, with the glyph for Aries at the beginning of spring representing the sprouting ear of corn, followed by Taurus the Bull as the spring mating season arrives. Leo ruled by the Sun indicates the heat of midsummer, and is followed by a harvest-deity holding a sheaf of corn in the season of harvest. Libra the Balance expresses the weighing out of the autumn harvest, then Scorpio, astrologically linked to dying and rebirth, represents the time of the year when death and decay is seen in nature as winter begins.[11]

It is thus appropriate that the Tropical zodiac started out life as a calendar system: in the fifth century BC, the twelve zodiac signs envisaged as months

---

10 Liba Chaia Taub, *Ptolemy's Universe* Illinois 1993 p.139. Claudius Ptolemy (c.100-176 A.D.) made his own observations at Alexandria during 127-141 A.D., when the two zodiacs were a degree away from coinciding. Books VII and VIII of his Almagest deal with the fixed stars.
11 Rupert Gleadow, The Origin of the Zodiac 1968 p.28.

were anchored to the solstices and equinoxes, by a Greek astronomer Euctemon. Only later did it become a framework for celestial measurement.

In Pliny's multi-volume 'History of Nature', composed around 60 AD, there is much discussion of agricultural customs of the ancient Romans based upon the lunar month. It was a scientific treatise of the ancient world, making his view of the zodiac of special interest. Was it tropical or sidereal? It's quite hard to answer that, because all he says is that the Spring and Autumn Equinoxes are located at eight degrees of their respective signs, i.e. Aries and Libra. In his world, in which the Earth did not move, these were pivotal points on which the Four Seasons revolved.

There was no easy way of comprehending how these could move against the stars. So Pliny used a tradition, a memory, from the earliest time when the zodiac was first formed five centuries earlier! He was using a memory of the star-zodiac, as long before him had the Vernal Point embedded at around eight degrees of Aries.[12] Ptolemy effectively unlinked the Zodiac from the stars and connected it to the Four Seasons. His zodiac was identical with the ancient Babylonian zodiac at that time: Ptolemy's zodiac was sidereal, in that his section in Tetrabiblos on the zodiac is entitled "The influence of the Fixed Stars", and refers throughout to the individual stars which comprise the zodiac images.

One end of Taurus is different from the other by virtue of its different star-clusters. In like manner he refers to the influence of various extra-zodiacal constellations on man.[13] But, his zodiac was firmly linked to the seasons of the year, i.e. it was also tropical! Indeed, he explained their passage with reference to the Sun's motion through the signs (there were no southern hemisphere astrologers around to dispute this). Uniquely, the definition of his zodiac had components of both the tropical and sidereal systems: he appeared as a bit of a Janus-figure who was able to face both ways, backwards to the old Sidereal scheme and forwards to the tropical zodiac that would develop, unlinked from the stars.

According to Fagan, astrologers/astronomers following Ptolemy were by no means generally inclined to accept this fixing of the Vernal Point at zero degrees of Aries:

> as late as the 5th century A.D. the fellow-countrymen and successors of Claudius Ptolemy still used the Sidereal Zodiac...we must conclude that the moving (i.e. Tropical) Zodiac was not in

---

12 Pliny *Historia Naturalis* XVIII Ch.59, p.221; Otto Neugebauer, 'A History of Ancient Mathematical Astronomy', II, 1975 p.594.
13 Claudius Ptolemy, Tetrabiblos Loeb Clasical Library, 1940 pp.201-203.

common use in the Near East during the first 500 years of the Christian Era.[14]

Ptolemy may have proposed measuring longitudes from the vernal point, but this theoretical suggestion was far from being accepted by practicing cosmologists/ astrologers. Their respect for the Chaldean Magi meant that they held onto the sidereal tradition.

It was only with the advent of the 'Dark Ages' in the 5th-6th centuries A.D. that this tradition ended and Ptolemy's two books were taken by Arabs as their only or chief link with this past tradition. Ptolemy's fixing the start of the zodiac at a position in space defined by a point in time of the year's cycle, the equinoctial point at March 21st, became henceforth the only zodiac they knew about. Thus the Sidereal zodiac vanished from the West.

In the East, "In its original form the zodiac in India was probably the zodiac used by Greek Astrologers, which, owing to the spread of astrology, became transmitted to India in the 2nd century A.D."[15] For centuries, India has used a star-zodiac defined by putting Spica at 30° of sidereal Virgo. This is as we have seen much the same reference as was used by astrologers of the Hellenistic world.

It was not until the Arabs translated Ptolemy that they came to define their zodiac solely by the Vernal Point, so that the Tropical zodiac came into use. The Sidereal zodiac was obliterated by the Dark Ages with the demise of Greek culture. A clearly tropical reference for a zodiac may not have appeared until the fifth or sixth century A.D. when the Arabs took over Ptolemy's work, after the burning of the library of Alexandria helped to finalise the loss of all connection to the original star-zodiac.

The Sidereal zodiac was in use for a thousand years, greatly ignored by historians. Still, histories of this subject usually ignore the star-zodiac completely, moving from the unequal constellations of the ecliptic straight to the modern tropical zodiac.

The stages of evolution of these zodiac-forms indicates a movement away from direct experience of the night sky, and towards abstract calculation. From seeing the Moon against the constellations of the night sky, the progression was towards abstruse calculation, based on something quite invisible, namely the axis of the Vernal Point/autumn equinox. As the zodiac of the astrologers slowly crept away from the star-constellations which had given it birth, astrology retained no memory of the star-zodiac of the Sumerians, Chaldeans and Egyptians: the child forgot the existence of its parent.

---

14 Cyril Fagan, Zodiacs Old and New 1951, p.18.
15 Quoted from Powell and Treadgold, ref (5), p.17

The tropical and Sidereal zodiacs have presently moved nearly 25° apart, moving one degree per human lifetime (72 years). In our century, the vagueness over the star-zodiac is shown by a huge range of estimates over when the Age of Aquarius will dawn.[16] The Vernal Point, now at almost 25 degrees of sidereal Pisces, will emerge from this sign in three and a half centuries from now.

20th Century Boundaries

Astronomers in the twentieth century have (rather shockingly) terminated the Greek tradition of twelve zodiac constellations - of what we rather loosely call twelve zodiac constellations, but should rather call ecliptic constellations, i.e. as crossing over the centre of the zodiac. The Greeks always had more than twelve constellations of the zodiac, say fourteen including the fin of Cetus the whale, which do not reach the centre viz. the ecliptic.

In 1928 the International Astronomical Union decided to place boundaries around the constellations, and impose a structure upon the teeming multitudes of stars: every star would belong to a constellation.[17] A problem arose with a traditionally overlapping zone of the ecliptic, where the foot of Ophiucus the serpent-bearer stepped onto the Scorpion. Traditionally this had always been viewed as belonging to the zodiac constellation of the Scorpion, so that all textbooks alluded to twelve constellations of the zodiac - not 13!

If you check out any astronomy books you have, they will probably allude to the twelve zodiac constellations: the habit of millennia is not easily altered. Modern astronomers like to distance themselves from astrologers, so they took a rather shocking opportunity, whereby they were able to break up the millennia-old concordance between twelve constellations lying on the ecliptic, and the twelve signs which they mirrored. They thereby established thirteen constellations on the ecliptic, an unheard-of thing. On the ecliptic, the line along which celestial longitude is measured, the 'thirteenth' zodiacal constellation 'Ophiuchus' was assigned twenty degrees by the astronomers, leaving only six degrees to the Scorpion.

The Anthroposophical star-calendar was drawn up in the 1930s (chiefly by Elisabeth Vreede),[18] loosely based on the newly-established I.A.U.

---

16 For a skeptical review, see *The Gemini Syndrome, A scientific Evaluation of Astrology* by Culver and Ianna New York 1979, 1984, Ch.6.

17 The boundaries were published in 1930 by the IAU, and one sometimes hears this year cited as the date for the IAU fixing of boundaries.

18 The *Sternkalendar* was developed in 1936-42 by the Mathematical-astronomical section of the Goethenaum, started by Elizabeth Vreede from her Calendar begun in 1929. (There is a synchrony between this event in Switzerland and the IAU fixing the constellation-boundaries in Belgium in 1928-30). After the big split in the Anthroposophical movement

boundaries, but wanting no truck with its thirteenth constellation. Anthroposophical books tend to claim merely to be using the modern astronomical (I.A.U.) divisions as the basis for their calendar, rather forgetting these key adjustments.

If we compare the lengths of four adjacent zodiac constellations as the I.A.U. defined them in 1928, in celestial longitude to the nearest degree, with the same constellation divisions as used in the Thun calendar, as taken from the Sternkalendar of the Anthroposophists,[19] we see that the Anthroposophists reduced the already short constellation of Libra to a mere 18°. Measuring these four constellations along the ecliptic:

	Sagittarius	/	Ophiuchus	/	'Scorpio'	/	Libra
I.A.U. boundaries:	33°	/	19°	/	6°	/	23°
Sternkalendar:	30°	/	-	/	31°	/	18°
Difference:	-3°	/		/	+25°	/	-5°

In 1980 a sidereal Ephemeris was published in the US, based upon the Fagan-Bradley definition of Aldebaran's position.[20]

Historically, the position of the sidereal zodiac was and is in dispute over a degree or so but not more: viz, whether Spica be taken as 29° or 30° of sidereal Virgo.[21] There are some respects in which such a difference could be significant, however it has no importance as far as our calendar's sidereal ingresses are concerned. Readers with real stamina will by now comprehend the four different celestial reference systems discussed so far, and be ready for one more which is the last to be here reviewed. Summarizing the progress so far:

Vreede was thrown out in '35 and for a few years two different and incompatible 'Kalendars' appeared, with different constellation-boundaries. In 1936 the 'official' calendar had the Sun enter Aries on the 21st April while that of Vreede had it enter on the 18th, a four-degree difference; the current *Sternkalendar* has it enter on the 19th - Vreede's view! Likewise the Sun's entry into Scorpius was in 1936 given as 12th September by the official calendar, 15th by Vreede's. The book 'Anthroposophy and Astronomy' by Elizabeth Vreede (2001) makes, curiously enough, no mention of this topic.

19 These I.A.U. figures were taken from Powell and Treadgold, ref. (53). The Ophiucus size was read off from Norton's Star Atlas. Powell and Treadgold also gave the 'anthroposophical' constellations to the nearest degree.

20 Neil Michelson's *The American Sidereal Ephemeris 1976-2000*, Astro Computing Services San Diego 1980 contains an (anonymous) Foreword outlining the recovery of the ancient Chaldean zodiac in modern times, citing the Huber article. It puts Aldebaran at 15°03' of sidereal Taurus and Antares as 15°01' of Scorpio. Best & Kollerstrom used this ephemeris in Planting by the Moon, ACS 1982-4.

21 *Astronomy Before the Telescope*, Ed. Christopher Walker, 1996, argued for Pollux having been placed at 30° of Gemini, as would put Spica at around 30° of Virgo (Ch. on 'Astronomy & Astrology in Mesopotamia'). This differs by one degree from the ref. (20) position.

I Sidereal zodiac - the original Chaldean/Egyptian scheme, in use for two and a half millennia in the ancient world around the Mediterranean and then in India, advocated in 1980 by the Best & Kollerstrom 'Planting by the Moon;'

II Tropical zodiac - that of the astrologers from about the fifth century AD onwards, long used in rustic ephemerides. I and II differ presently in that one is rotated by nearly 25° against the other;

III   The 1928 IAU astronomical boundaries, with thirteen zodiacal constellations;

IV   The twelve ecliptic constellations of the Anthroposophists, derived from III by sundry modifications, used in Biodynamic calendars.

Paul Platt's Version

Our final framework is that of the American Paul Platt, who derived twelve sidereal constellation boundaries from his own (claimed) experience, and composed bulky volumes on the subject entitled 'The Qualities of Time.' The divisions obtained by this bold enterprise[22] were used by the Kimberton Hills calendar for 1988 and 1989. Due to complaints by a small but influential minority, this was withdrawn.

The Platt alternative has some strong claims to be taken seriously, and not merely for the reasons for which Platt developed his scheme: primarily, it gives a roughly equal balance of the four elements, whereby it differs markedly from the Biodynamic scheme. The latter, as we have seen, suffers from a grave imbalance between the Air and Earth elements.

The US 'Kimberton Hills' calendar is largely an echo of the Thun calendar but presented differently. I met Sherry Wildfeuer at Kimberton village, who produces the yearly Kimberton Hills calendar. The subject of the equal-interval sidereal framework arose, and her view of the divisions was based on her experienced perception of the constellations in the night sky, something not normally available to today's city-dweller. Consulting the star-zodiac diagram prepared by Robert Powell, she explained why this was not acceptable to her. Leo the Lion lost his head, for a start. This seemed a forceful argument. Was it possible, she suggested, that once upon a time the

---

22 A shorter work by Paul Platt is *The Moon's Rhythms at Work*, a guide for parents, teachers, therapists (and anyone else) to observing the influence of the moon's rhythms upon children's and their own behavior through the course of the year NY, 1991.

divisions had been equal, but that in the course of time things had changed a bit?

After that, I met Paul Platt at a Washington conference of astrologers. Could he draw in his constellation boundaries, I asked him, on the sidereal grid? He couldn't, he explained, as his experience of these things was solely as a process in time. He knew his twelve divisions as tropical longitudes only, and was unaware of how they pertained to the traditional constellation-images; or so he told me and I have no reason to doubt him.

He soon converted these, however, giving his own divisions. Leo keeps his head ... but Pisces loses one of the fish, so that the huge but dim constellation of Pisces is axed. Libra acquires a most unclassical expanse, while the adjacent constellation Virgo becomes rather slim, and loses its bright star Spica! For her to lose her brightest star, always traditionally linked with fertility and good fortune because of this image, seems rather shocking. Ancient questions of how these sky-images were derived are thereby raised. Mr Platt had no comment, explaining that his divisions had not been derived from such considerations. There were two sources for his notions: his experience of lunar transits, which passed by every two to three days, and that for solar transits which he was for a period able to experience in the course of each year.

Paul Platt asked me if I would reanalyze the Bishop radish data to see how his boundary divisions scored. I had already used them for three different frameworks, no doubt taxing the reader's patience considerably; however I did this and they weren't quite as good as the equal-interval divisions. Let us hope that Mr Platt's experiential new approach can stimulate constructive debate over these matters. His constellation boundaries appear as greatly preferable to those used by the Biodynamic calendars.

In 'Planting by the Moon' written in 1982 we claimed that twelve irregular constellations of the zodiac had never been used by cosmologists for their concepts of influence. That may have been true then, but is no longer the case, with an alarming trend amongst astrologers to use them for their arguments about when Aquarius is going to 'dawn.' Some dawnings of the Age of Aquarius put it around eight hundred years' hence, due to the large size of the constellation of Pisces.[23] On the whole astrologers ignore the starry background behind their zodiac, but from time to time they become uneasily aware that they need some notion of a starry boundary to define this hoped-for event. Books about the astrological ages give them all

---

23 *The Astrology of the Macrocosm* Ed. J. McEvans, Llewellyn 1990, Nicholas Campion 'The Age of Aquarius,' pp.214-7. He argued that the constellations of the fishes should be used to gauge precession. Since there were (in his view) many possible starting-points for a sidereal zodiac, which was therefore uncertain, the patterns of the stellar constellations were preferable for determining the zodiac 'ages'.

different durations, with diagrams of the unequal constellations looking somewhat similar to those of the Biodynamic calendars. Nowadays the sidereal zodiac is alive and well, with the spread of Indian astrology. Let us hope that a renaissance of interest in this primordial zodiac will begin as its botanical significance begins to be appreciated.[24]

Goats' milk and the fire-trigon

A woman milks her goats each morning. In this simple, rustic scene there was something new under the sun: each day she was measuring the net yield of milk from the goats, then making butter from it, and then weighing its yield.[25] No respectable scientist would perform an experiment as simple as this one. The woman was Maria Thun, and the fabric of Time was being investigated.

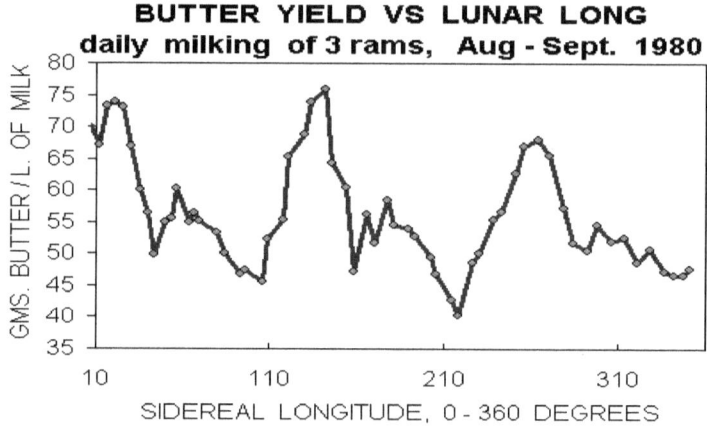

Figure 7.3 Yield of butter in grams per litre, obtained from daily goats' milk over sixty days, plotted against lunar sidereal longitude 0-360° at milking. The line shows a 3-point moving average. The same was plotted for merely 0-120°, the line being a 14-point moving average.

Daily, over two months, did this show any lunar phase effect, or any perigee effect? One might expect so, but they were not evident (in my view).[26] Instead, the peaks corresponded to the three warmth-constellations. One might expect that such a daily experiment would enable comparison of

---

24 Robert Powell, *The History of the Zodiac*, (Sophia Academic Press, CA, 2007) alludes to my work on pp.7 & 81-82. Then, his *The Astrological Revolution* (Lindisfarne books, 2010) has an Appendix 4 'Planting by the Moon,' giving the argument for an equal-interval zodiac framework for use with growing crops
25 M. Thun, *Milch und Milchverarbeitung* 1991 (self-published) p.22.
26 Her 2002 Calendar avers that perigee days 'are nearly always unfavorable to milk processing, including making yoghurt.' No data from her milk-book supports this.

constellation-boundaries with equal-interval signs; however it happens that the three warmth-constellations are not so different from the corresponding sidereal 'signs'.

We should aspire in a Goethean sense to find experiments sufficiently simple that they can ask, as it were, the primal questions of Mother Nature, so as not to put Her upon the rack of interrogation as modern scientists like to do. Whatever it is that the Ram, the Lion and the Archer have in common, whatever words with which we seek to describe the quality of this sky-triangle, first conceived in ancient Babylon: the goats of Thun were responding to it. They gave the best butter-producing milk, i.e. the best quality milk, as the Moon thrice-monthly was activated by that pattern.

Readers with goats might consider replicating this experiment, because until someone does so its status has to remain uncertain. The person who does the milking ought not to be the same as the person who collates the data, and the problem becomes more acute if the one person involved has propounded the theory, as supposedly showed up in the data. Science is inherently a public enterprise, a shared endeavor: however we often run into this problem in what one might call astrological-type research: that the investigator is to a degree isolated and we lack witnesses as would have been helpful.

The Ram constellation does not score any higher than the Lion, also associated with the warmth/fire element: it is a triune energy manifesting, not that of an individual sign or constellation.

On one interpretation the fire or warmth-element should be rather dry, and the water-element be more associated with milk. The four-element theory as we have dealt with it up until now gives us no indication on this matter.

BUTTER, SIDEREAL TRIGONS
AUGUST-SEPTEMBER 1980

Fig 7.4 Butter yields plotted by lunar longitude at milking time, over 0 - 120° to show the Four Element zodiac trigons. The yield of butter per litre of goats' milk, daily over two months, is plotted by lunar sidereal longitude. No-one predicted, we should note, that the peaks would be in the fire-trigon, rather than, say, the water-trigon.

To give an idea of how such data can be processed, let's divide the data to group the yields from each sidereal Moon-sign together with others of the same element. The resulting graph (Figure 8.4) spans 120° or one-third of the zodiac.

The data-points are given as deviations from a mean, and a trend-line goes through the data (a 'moving average'). If we express these four element-groups as a bar-chart (not shown), then the warmth-trigon milking-times would give around thirty percent more butter from the milk each day, than do the other trigons.

No perigee-effect was present in this data, i.e. I found yields to be unrelated to lunar distance.

Tasting the Wine

"How good is this wine?" That is a quintessentially qualitative question. Any reply involving number will be absurd. The wine connoisseur replies "Grown on the sunny side of the hill", or "It has a jagged edge," as they do. A little book by Maria and Matthius Thun appears, from Floris Press, in September 2009, 'When Wine Tastes Best,'[27] and suddenly the BBC, *The Guardian* and chains of supermarket stores are discussing, for the first time ever - the Star-Zodiac trigons! And this concerns not the planting or grafting of the vine, but something far more immediate: the tasting of the wine, the popping open of the bottle. But on the other hand, it does seem appropriate, indeed delightful, that opposite ends of the zodiac may be involved.

As you'd expect, the book advised that Fruit-days accentuated flavor, while flower-days brought out aromatics. Thus half the zodiac was 'good' for wine and half bad. Root days had to be avoided for tasting wine – yuk! Once the book was published, headlines like 'Supermarket chains check lunar calendar before inviting critics to drink' appeared. Tesco revealed that it had already been using this method (tasting in the fire-trigon) for the previous two years, but had kept quiet about it.

Let's quote The Guardian report:

> The idea that the taste of wine changes with the lunar calendar is gaining credibility among the UK's major retailers, who believe

---

[27] The Guardian 18.4.09 'Tesco and supermarket rivals go for wine tasting by moonlight'. *When Wine Tastes Best*, by Maria and Matthias Thun, is published at £3.99 by Floris Books.

the day, and even hour, on which wine is drunk alters its taste. Tesco and its rival Marks & Spencer, which sell about a third of all wine drunk in Britain, now invite critics to taste their ranges only at times when the Biodynamic calendar suggests they will show at their best. Tesco has used the calendar for more than two years to decide on times for its thrice-yearly critics' tastings, but has not shared its belief with customers for fear it will add yet more mystique to wine.

Marks and Spencer was first, we were startled to learn:

Marks & Spencer was the first big British retailer to hold its tastings on the most favorable fruit days four years ago, along with the small burgundy specialist Howard Ripley.- The Times, 22.8.09.

So, four years of testing this theory had already been going on the UK! We quote three witnesses from The Guardian's report:

Our first choice is a fruit day," said Pierpaolo Petrassi, Tesco's senior product development manager. "We seek to avoid root and leaf days. It may be a little step beyond what consumers can comprehend. We have so many other things to educate consumers about ... We don't want to make it more complicated.

Jo Ahearne of Marks & Spencer has said she was 'completely blown away' by the results of tastings carried out on root and flower days. She told decanter.com that over a period of several years the tasting team found noticeable and consistent differences between the same wines tasted on different days" 'We would taste 140 wines on one day and find no faults, then 40% of the same wines, tasted the next day, would be slightly duller.' Consulting the Biodynamic calendar, the team ascertained the 'off' days were root days. 'I was cynical about it,' Ahearne said. 'But now I'm convinced.'

The Guardian tested the theory this week and tasted the same wines on Tuesday evening, a leaf day, then again on Thursday evening, a fruit day. Five out of seven bottles showed a marked improvement. "I was skeptical but I think the evidence was overwhelming," said David Motion, the London wine merchant who hosted the tasting. "I live in the city and don't think much about nature but it is clear it has an influence. The cosmos is forcing its way in." Mr Motion added, "It wasn't that the wine tasted bad on the Tuesday but it was much more expressive on the Thursday. It was more exuberant and on-song. It was like the heavens opened, the clouds parted and the wine just expressed itself." The trial solved his long-standing puzzlement at why the same wine could taste so much better on certain days. From now on, he says, his wine shop in north London will only hold tasting sessions on fruit days.

These are business people expressing their views here, in a most forthright manner. This has to be a quantum leap forward in constellation research! And if I may say so it is quite alchemically meaningful, concerning the fiery essence of wine. People are warmed by drinking alcohol. 'In vinum veritas' went the old Roman motto, as if that warming of the inner being brought out some heart-sharing between drinkers. Pop open the bottle as the Moon traverses Leo, the majestic lion of the heavens, maybe as it goes across Cor Leonis, Heart of the Lion ('Regulus'). Also if the act of drinking somewhat inhibits rational thought, that may help these top London wine-tasters to feel this connection directly.

Of especial interest in this argument, is the way nothing is found to be more important than the human experience. All our lives we've heard scientists teaching our civilization to downgrade that which is 'merely subjective,' whereas here inescapably it is the only thing that matters. Arguably, indeed, it is not saying anything about the objective world: merely our experience of taste. Well, there is a bit more to it than that, because it is the fire-element within the drinker that is kindled, and a zodiacal context is here being found for that: Vettius Valens in Syria in the 2nd century AD who stuck the fire-trigon into the zodiac would no doubt have been pleased to hear this!

In 2010 the International Wine Fair in East London led with a plug of the new Thun theory:

> For the first time in over a decade, the show falls on the near perfect combination of tasting days in the Biodynamic calendar ... The Top 100 will incorporate a Biodynamic tasting booklet, allowing visitors the opportunity to compare their tasting notes across the three Biodynamic days. Key Biodynamic exhibitors will also be organizing on-stand activity to highlight the influence of the calendar.

Most of its website was dedicated to the new concept of the 'Biodynamic tasting days.' While welcoming this new twist to the Thun theory, one notes a slight elemental problem, whereby Air (for the 'flower days') falls opposite in its zodiac position to Fire (fruit-days).

The element-sequence, we remind ourselves, is Earth, Fire, Water and then Air. Ought not the 'bad' days to be four to five days away from the 'good' days in this nine-day cycle, as generally seen in the crop-yield experiments? But here, they are found to be adjacent.

The Air and Fire trigons are opposite each other i.e totally out of phase. But, if the element of Air (flower-days) is indeed better for the aroma of the wine, the equal-interval star-zodiac gives connoisseurs 50% more time to enjoy this, than does use of the Biodynamic constellations!

8. Planets & Perennials

... the powerful grace that lies In herbs, plants, stones, and their true qualities.

- Friar Lawrence, 'Romeo and Juliet,' Act II.

'Planting by the Moon' adopted a simplistic approach to planetary aspects, not least for reasons of space, citing only the Moon-Saturn aspects of conjunction, opposition and trine. Saturn, we roundly affirmed, was the ancient Roman god of agriculture, 'Chronos,' and therefore the Saturn-Moon aspects were good times for sowing perennials and trees. For planting trees and perennial crops, 'Planting by the Moon' advised that finding a correct planetary aspect was more important than the monthly rhythm of the elements, and recommended these Saturn-Moon aspects. Saturn's sickle originally had a more rustic meaning before it came to denote the limitations of Time.

The modern Biodynamic calendars became during the 1980s loaded up with planetary aspects: it appeared that someone believes they could detect their effects! In both the Thun and Kimberton Hills calendars one finds the recommendation that a planetary trine (a 120° angle between two planets) can define the dominant element, overriding the Moon-sign. This is because the two planets will both be occupying a sign of the same element. That is interesting, but raises the problem of two planets which can stay in trine aspect for long periods, years in the case of the outer planets.[274] However, it is nice to see lunar gardening calendars agreeing with each other.

Some remarkable research concerning planetary rulerships has been appearing from the mathematician Lawrence Edwards. After studying the shape of buds for many years, he has claimed that their morphology responds to the ruling planet in relation to the Moon. He has kindly made available his data and this chapter will comment upon this.

The Traditional View

> To the better furthering of the gardener's travails, he ought afore to consider, that the Garden earth be apt and good, wel turned in with dung, at a due time of the year, in the increase of the Moon, she occupying an apt place in the Zodiack, in agreeable aspect of Saturn, & well-placed in

---

[274] The Kimberton Calendar gives all trines, including outer planets; Thun, *Working with the Stars* 1982, pp.10-12, 'Trines'; also Thun 1991, p18. The Thun calendar does not give celestial aspects as such, it merely dispenses advice based upon them. In particular, trines are invoked as affecting the sidereal element-effect.

the sight of heaven ... for otherwise his care and pains bestowed about the seeds and plants, nothing availeth the Garden.

- Thomas Hill, The Gardener's Labyrinth, 1577, Ch. 21.

The sixteenth-century work on gardening lore, The Gardener's Labyrinth, explained about sextile aspects between Saturn and the Moon: "it is then commended to labour the earth, sow, and plant;" whereas during the square aspect between these two, it was "denied utterly to deal in such matters." The trine was also approved, but the opposition was not. Much the same advice featured in the early-seventeenth-century British work 'The Whole Art of Husbandry';[275] this was the Renaissance vision, in which Earth and Sky worked together in concordance, to produce a pleasant garden. It is our business to reaffirm its validity, and its practical utility.

Concerning celestial aspects, we may seek further guidance in a work of traditional astrology, as it contained a useful section on agriculture: 'Dariotus Redivivus, a Brief Guide to the Judgement of the Stars,' composed in the early seventeenth-century England. This gave as its first principle that:

> [the farmer] ... ought to have a special respect to the state and condition of Saturn, that he be not...afflicted, because he hath chief dominion over husbandry and the commodities of the Earth; let him therefore (if you can so fit it) be in good aspect with Jupiter or Venus...not retrograde, and in good aspect to the Moon.[276]

For planting crops, the general comment was:

> Plant what you intend, the Moon being either in conjunction, sextile or trine of Saturn, free from the body or beams of Mars ... with aspect of Jupiter or Venus. *

These quotes show the traditional notion of Jupiter and Venus as benefice planets, while Mars is a malefic: the warrior-principle of Mars is to be kept out of the garden, while the traditionally feminine 'planets' Moon and Venus, associated with fecundity, are needed. The opposition and square aspects, involving 90°and 180° angles, were not recommended. Traditionally, these would be associated with stress, firmness of structure etc, which might not assist a growing thing, while trine and sextile aspects (120° and 60°) are associated with the harmonious flow of energy. I told an elderly astrologer that Biodynamic gardeners were keen on planting things at Saturn-Moon oppositions and he was quite shocked. This did not appear

---

275 G. Markham, *The Whole Art of Husbandry,* 1631 (trans. of a continental work, originally published in the 1570s).
276 Dariotus 1653, *Certain Rules for Husbandry*, pp.180-2.

to him as being at all a fertile configuration. It was not so long ago that this was the sole celestial aspect to be found in their calendars.

An experiment by Maria Thun with grain harvests concerned Mars. She sowed varieties of wheat in the Autumn of 1993, which next summer were harvested at 'twelve different times'. It was found that "The lowest weights with all varieties came from seed harvested during an occultation of Mars the previous year."[277] This should give food for thought for farmers concerned to improve their seed stock, concerning when to harvest the seed. This confirms the traditional image of Mars, as being dry and hot and not of much help in the garden.

For grafting and pruning the waning Moon was recommended by Dariotus, in aspect to the benefices, advice passed down since time immemorial:

> Gather Apples and other fruits, herbs and flowers to have them faire and beautiful, when the Moon is towards the full; but to keep and reserve in store, gather them dry, the Moon being in her decrease and configured to Jupiter or Venus.

It is of interest to compare this traditional view with that expressed by a US publication, 'Gardening Success with Lunar Aspects' by Adele Barger.[278] This cited the sextile and trine as beneficial aspects while the square and opposition were detrimental, and the conjunction could be either, depending upon circumstances. A narrow orb was advocated for lunar aspects of just three degrees, which means that sowing or whatever should be performed within six hours of the celestial event in question. Sun-Moon aspects are especially recommended, such as the trine. Moon-Venus aspects are recommended for all ornamental crops, which again seems logical. Saturn aspects bestow hardiness, while Mars aspects bestowed fast growth, the latter being quite untraditional.

She used all of the planets including outer ones for the aspects, viewing none as inherently malefic. A computer program could readily assess optimal times on this basis, whereas it would be difficult and tedious for a gardener.For Moon-Venus aspects, Ms Barger advised that the conjunction was "excellent", the trine "good" and the sextile "slightly good." The Moon and Venus are traditionally feminine planets, and she felt that their energies

---

277 Thun, *Working with the Stars*, 1994, p.9. Quite generally, Thun claims to have found through experiment that Moon-planet occultations tend to undermine seed quality, whereas the oppositions, two weeks later, are generally beneficial for the crop: 'Soil, moon and Stars, a Biodynamic workshop with Maria Thun,' *Star and Furrow,* Winter 2001/2, pp.13-15.

278 A.E. Barger, 1977, self-published; 2nd Edn. 1983, AFA. Ms Barger also recommends the traditional rules whereby above-ground crops are to be sown in the waxing half of the lunar cycle, and bulbs and root crops in the first week after the Full Moon; she found however that nothing benefitted from being sown in the last quarter. Her zodiac rules were the usual, that the water signs were fertile and Libra best from flowers etc, none of which are here endorsed. This booklet gave eight years of daily gardening advice in a little over fifty pages.

blended well at conjunctions; whereas in the case of Saturn "The conjunction is unfavorable because the Moon and Saturn are of such different natures that their energies fail to blend well."

This is all rather impressive, and Ms Barger would appear to be about the only person in the modern world who has attempted to translate astrological notions of celestial aspects into a rural context. However, there is a major logical contradiction which she appears not to notice: in her scheme, the Full Moon should be detrimental, as being an opposition aspect. She plainly cannot take this view, and in fact advocates the sowing of fast-germinating seeds within three to four days of the Full Moon.

This is quite an interesting, and reasonable, modification of Kolisko's recommendations made in the 1930s. A celestial aspect may be viewed in terms of two halves, one of which is 'approaching' or 'applying' and the other which is 'separating,' divided by the moment in time when the aspect reaches exactitude. Ms Barger endorsed solely the first half, i.e. its 'applying' phase as being potent for the various gardening activities. In this view, a lunar trine would give six hours for optimal sowing, ending as the aspect chimed: 'The instant the heavenly bodies begin to separate from their aspect, the energy is dissipating and no longer in effect.' This is a rather extreme view. The trine might form in the middle of the night, when it would be useless.

However, an emphasis on the approaching aspect finds support from tradition: we describe below how Culpeper's 'Herbal' instructed that the Moon should 'apply' to whatever planet-aspect was chosen, which would refer to the two planets coming into their aspect, as opposed to their separation.[279]

Rulerships Old and New

For perennial crops and trees, a traditional notion of when to plant can be derived from the notion of 'rulership,' whereby for example the rose is said to be ruled by Venus. This would be relevant, for example, if a farmer wished to plant an orchard. What Dr Steiner said about the matter in 1924 was:

> If someone wishes to plant an oak, it is of no little importance whether or no he has a good knowledge of the periods of Mars; for an oak, rightly planted in the proper Mars-period, will thrive differently from one that is planted in the Earth thoughtlessly, just when it happens to suit.[280]

---

279 Medical astrologer Gary Price kindly explained the significance of Culpeper's astrological remarks concerning the picking of herbs.
280 Rudolf Steiner *Agriculture* lectures 1924, p.27.

A problem here is that oak trees have always been traditionally viewed as ruled by Jupiter. After all, the oak groves were sacred to Zeus. An oak-tree supposedly nourishes more different forms of bird and animal life than other trees, and could express Jupiter's expansive nature. These traditional rulerships may be found in Culpeper's ever-popular 'Complete Herbal.' The 'Book of Rulerships' by Dr Lee Lehman conveniently summarizes rulerships as they were given by ancient sources and gives no suggestion of a Mars rulership for the oak anywhere in past history.[281]

In his lectures on 'Biodynamics,' Podolinsky discussed planetary rulerships, citing the oak as being under the dominion of Jupiter. 'It carries the Jupiter power of being' he averred,[282] and described in what ways it seemed to him to have such a quality. Personally I tend to experience the gnarled and knotty oak as 'ruled' by Mars if anything; however there does exist a major problem, in terms of historic continuity, if one entertains the notion of 'rulership' for trees.

A tree more free of controversy as regards its rulership is the beech, which is given by Culpeper as coming under Saturn. Is not the quality of a glade of beech trees rather Saturnine? It has a serene sense of calm and old age, and hangs onto its dead leaves from the previous year right into April, which appears to be a 'signature' of its inner being. Dr Lehman cited only a rulership of Jupiter for the beech, as given by the English astrologer William Lilly (she advised the writer that she had not noticed the Culpeper reference).[283]

Culpeper gave the sunflower as ruled by the Sun, and poets have extolled its close connection with the Sun, shown by the way a field of sunflowers even follows the Sun's daily course as their 'necks' turn. But, according to Rudolf Steiner, it was ruled by Jupiter. Here we may not have so great a problem, as there are no mediaeval sources for this rulership. It was brought over from Mexico in 1659.[284] Having been introduced in more recent times, it may have been assigned a solar rulership merely from its appearance, which is not quite the relevant factor.

Kolisko originated the notion that planetary rulerships were testable by experiment. It was not necessary to rely upon tradition, she averred, this having become rather fragmented. For the sunflower seeds, Kolisko found that the seeds grown in a tin-lined container germinated better than others, tin being the Jupiter-metal. Other experiments of hers, including use of dilute metal salt solutions, and germinations during Jupiter-Moon

---

281 Lee Lehman *The Book of Rulerships* Whitford Press USA, 1992. Dr Lehman's sources ranged from antiquity up to the seventeenth century.
282 Alex Podolinsky, 1990, p.164.
283 Nicholas Culpeper, *Culpeper's Complete Herbal*, Foulshams, p.48.
284 E. and L. Kolisko, 1940, 1978, pp.52-54.

conjunctions and oppositions, tended to confirm this rulership. As a result of these experiments, the sunflower appears as the sole crop whose planetary rulership has been well investigated in modern times - or at least, up until the 1980s, when a new approach was initiated by Lawrence Edwards. A summary of the three stages of Kolisko's procedure may be quoted:

> (Kolisko) made dishes out of the various metals [gold, silver, copper, iron, tin and lead], filled them with rainwater and placed seeds from the same sunflower in each. She discovered that the seeds in the tin dish started germinating first after only 24 hours, twice as fast as mercury and over three times as fast as lead."
>
> In another series of experiments Lilly treated the sunflower seeds with the seven different metallic salts, from the 1st to the 60th potency, over 2000 plants in one experiment alone. Again, the seeds treated with tin were taller, stronger and healthier than the others and, in fact, the sunflowers treated with the Sun's metal gold grew quite poorly.
>
> "To take this study a step further, Lilly also germinated seeds during the opposition of Jupiter with the Moon and during the conjunction of these two planets. Using tin chloride she discovered that the sunflowers grew much higher and better during the conjunction of Jupiter and the Moon, than during their opposition. Even the control plants in rainwater grew better during their conjunction.[285]

These experiments assumed a functional relationship between the planets and metals traditionally associated with them, in this case Jupiter/tin, which may strain the credulity of some readers.[286] Surely one would appreciate a packet of flavor-enhanced sunflower seeds, from sunflowers sown during some strong and harmonious Jupiter aspect? Would one not pay more for it?

The concept of rulership implies that some fairly tranquil mode of celestial energy, varying with planetary events in the heavens, works in a plant. The farmer may use it by planting his tree or whatever at an optimal time. The enhancement of power comes through a lunar aspect: conjunction, opposition or trine. Kolisko concluded that lunar conjunctions were optimal, and these come round once a month.

---

285 Alison Davidson, *Metal Power, The Soul Life of the Planets*, Borderland, California 1991, p.45.
286 See e.g. Rudolf Hauschka,*The Nature of Substance,* 1966, 1989; or, N.K., 'The Metal-Planet Relationship', Borderland, California, 1993.

For a farmer sowing a field of sunflowers, the sidereal rhythm remains relevant. In the fine words of Thun:

> Cultivated plants which do not become woody live in close relationship to this rhythm [the sidereal].[287]

One would sow in the Fire element if seed was required. Within this context, one could try lunar aspects to the Sun or Jupiter at sowing time, to see whether Kolisko's findings are confirmed. When gazing up at a sunflower, which can readily grow to being ten feet tall, one can wonder whether any special quality is there expressed.

Dr Georg Schmidt in Germany owns a tree-nursery and has been investigating sowing times in relation to subsequent growth, carrying on the work of his father Martin Schmidt. It is his experience over 13 years that the opposition, between the Moon and the planet supposed to be ruling it, is a better sowing time than the conjunction.[288] This is somewhat puzzling, as there is nothing much in traditional lore to say why the opposition should be preferable to the conjunction in this respect. This is slow work but let us hope we hear more about Schmidt's findings in due course. The planting of trees is a matter of the highest importance in the modern world. There is such a thing as the art of planting trees, which involves choosing the time which suits them, rather than that which suits us. If a new science of time can aid their growth then it deserves our fullest attention.[289]

The UK astrologer Wanda Sellar is also an aromatherapist, having written a textbook on the subject, and she gave a presentation about herb planetary rulerships in which the scent of each herb was especially emphasized. These were the ones she selected:

> **Lavender** = Mercury , **Rose** = Venus
>
> **Orange** = Sun , **Ginger** = Mars ,
>
> **Jasmine** = Jupiter , **Cypress** = Saturn

Lunar herbs (she explained) were less aromatic, however she selected oil of clary sage for the Moon.

---

[287] Thun, Working with the Stars, 1977, p.2.

[288] Georg Schmidt, 'Aufbau lebensfahiger Naturbereiche...', *Lebendige Erde* 1984, 2, pp.3-15; for a fine summary of his work, see Biodynamics, New Zealand 1989, p.138. His conclusion about oppositions being preferable to conjunctions is the converse of Kolisko's from sunflower growth, where she found that the conjunctions worked better.

[289] In *Planetary influences upon plants* 1984 (from the German) E.M. Kranach wrote: 'J. Schultz during the 1940s studied seed formation in relation to the 12-year Jupiter cycle. For red beech, the years of high seed yield coincided with specific constellations of Jupiter against the zodiac.' (p.3) I have little sympathy with Kranach's approach, which appears here to be assuming a Jupiter rulership of beech.

Improving the Wine

Fortunes are made in the wine industry, so it is hard to believe that wine growers nowadays fail to heed the astrologer's advice about the time of planting the vine, which would make so great a difference to the quality of the product. Ancient authorities were unanimous over the rulership of the vine: the Sun owns it. Its growth depends on having a sunny spot, and on the amount of sunlight over the season. It also seems to depend upon how many sunspots appear on the face of the Sun.

To quote the late Michel Gauquelin:

> According to the French Astronomical Bulletin, years in which the number of sun-spots is highest are great vintage years for Burgundy wines; in years with few sun-spots poor vintages are produced. The Swiss statistician A. Rima found similar results when he analyzed the production of Rhine wines for the past two hundred years.[290]

The sunspot maximum of 1990 was regarded as a classic wine-year. Other things are implied by the notion of a solar rulership in this context, related to the traditional association of the Sun with the heart, and the fiery qualities of Sol.

The words jolly, jovial and joy derive semantically from Jove, which deity-principle is linked mysteriously with a star in the sky, namely the planet Jupiter. For planting or grafting vines one should seek out a Sun-Jupiter or Sun-Venus aspect, Venus to bestow sweetness of taste upon the wine. The Sun moves one degree per day along the zodiac, so such an aspect lasts a few days. That would give, within a week or so, the date when the new cutting is placed into the soil. The Sun trines with Jupiter twice a year, and these events may not fall within the growing season. The ancients related the quality of wine in a given year to the twelve-year sojourn of Jupiter around the signs.[291]

Our 'Planting by the Moon' ignored lunar sextiles, these being the 60° aspects, as being too weak in nature: however for the slower-moving cycle between the Sun and Jupiter, they become significant. Being half of the angle of a trine, we can imagine their effect as somewhat comparable. A trine (120°) signifies the harmonious flow of energy, and such an aspect might turn up in the growing season. For a sweet wine, an aspect to Venus might be preferred, which effectively means a conjunction as no other

---

290 Michel Gauquelin, The Cosmic Clocks 1973, p.100.
291 *Agricultural Pursuits*, Trans. from Greek by the Rev. T. Owen, 1805, section XII. See also ref. (91)

aspects are formed: it remains too close to the Sun in the sky to make any of the necessary angles.

No-one I have asked has guessed correctly the traditional rulership of the vine, as given by the classical authorities. We saw how the 'Book of Rulerships' by Lee Lehman featured a group of plants over which the traditional sources were more or less unanimous (Dr Lehman has her doctorate in biology and is a well-known US astrologer) and the vine was one of those.[292]

One meets outright incredulity on this matter! Dr Zip Dobyns (she was a well-known US astrologer/psychologist) suggested that such a rulership could have changed over the ages, concurrent with changes in society. Well, perhaps it could. The question was raised as to whether it was really the grape vine to which such rulership pertained. A book on plant rulerships by Jean Eliott clearly affirmed this. In its solar rulership section it said:

> GRAPEVINE (Vitis Vinifera). Lilly, Culpeper. Either grows as a long-lived climber or in bush form for wine. Green/white flowers from early to midsummer. Grapes in Autumn. Under Vine in Herbal. Brought to Britain by the Romans.[293]

She summarized the solar qualities as follows:

> The Sun: Core essence, integrated conscious self, playful self, vitality. Play, children; palaces and mansions; day; gold. The Sun rules Leo. Colour: yellow, orange, gold.[294]

Growers skeptical of these matters are invited to plant a vine over an eclipse - or, at a Mars-aspect to the Sun, for example a square (90°), without any Moon-aspect being present. This should give a bitter-tasting beverage, indeed the vine itself may appear rather mangy. The reason for this is that the traditionally-understood qualities of the Sun and Mars are in themselves dry and fiery, and so taken together they would be rather barren.

If the event is fairly important for future prosperity, then I suggest planting/grafting the vine at noon on the chosen day. No-one is being asked to be up at sunrise! Theoretically, the rising of the Sun, in its exact aspect to Jupiter, is the most influential moment, but culmination, i.e. the Sun reaching its highest position, or possibly sunset, on the day of the aspect, will do well. Use of a key point in the diurnal cycle should enhance the

---

292 See Lehman, ref.(8).
293 Jean Elliott, *Plants and Planets, Astrological gardening with a Comprehensive guide to plants assigned to planets by 17th century astrologers Nicholas Culpeper & William Lilly*, 1996, Herts, pp.4,17. 'Hemp, Marijuana' is given as under Saturn, as it was used for rope-making and ropes had to be sturdy and reliable. The view expressed in certain quarters is that cannabis might be ruled by Neptune: however the outer planets are excluded from this rulership book.
294 Ibid, p.16.

celestial event selected. The important moments for establishing a vine are cutting the scion in the Autumn and then grafting it in the spring.

Maria Thun has been working with a French Biodynamics farm to improve its grape vines, for which it won a prize, and I heard that one could see the difference between the vines of that farm and its neighbors. About 1% of French wine is presently Biodynamic, and this figure is increasing. Let us note therefore that Thun's recommendations would not include any of the recommendations as regards celestial aspects here given. This is the second major respect in which she has not endorsed the researches of Kolisko. Thun's use of planets in agriculture does not seem to utilize the notion of rulership. For example, her recommendations as to when to sow sunflowers would not take account of Jupiter's position. [295]

The French winegrower Nicholas Joly has been deeply influenced by some visits made by Maria Thun to his vineyard in the early 1990s. His Coulée de Serrant vineyards at Savennières, Maine-et-Loire produces some of the most highly esteemed wines of France. Wine connoisseurs accept that the quality of wine from this long-established vineyard has improved following Joly's adoption of Biodynamic methods in the late 1980s. In wine the notion of quality reaches its apotheosis and becomes the supreme issue.

In his two books about Biodynamic wine,[296] Joly proves that he is able to find the words to discuss this matter – and is well aware from his experience of the pitfalls of the inorganic, chemical approach which the 'experts' have to offer. His discussion on how quality tests are used, especially the copper-chloride crystallization method, is of value. Wine-growers need to read these books, maybe together with the chapter on the subject in Thun's (English translation) 'Results from the BD sowing and planting Calendar.' London grocery shops often have a section on organic wines these days, and are happy to obtain BD wine for the customer: is it worth paying the extra, and is it true what they say about not getting that hangover the next morning?

Joly affirms the importance of 'trigon treatment' when working on the vine, i.e. choosing a day when the fire-trigon has the Moon or planets therein. The specific planets he recommends here are Mercury and Saturn,[297] of which one may be doubtful. Traditionally the relevant 'planets' would be the Sun or Jupiter, and for flavor one might add in where possible, Venus. There is room for further debate over this matter.[298] A traditional view would have Saturn giving the wine a dry and bitter taste – with quite

---

295 See interview with BD Wine-expert Monty Waldin, author of *Biodynamic Wine*.
296 Nicolas Joly, *Wine from Sky to Earth, Growing & appreciating Biodynamic wine*, 1999, Texas USA; 'What is Biodynamic Wine? The quality, the taste, the terroir', E Sussex, 2007.
297 Joly, Ibid 2007, p.100.
298 Joly, Ibid 1999, p.57 and 113.

a hangover! Joly advises to harvest the vine-grape, as well as sowing it, with the Moon or planets in the fire-trigon.

The grape-harvest usually takes place on his estate from late September to early November. That harvest-period sees the Sun move from the Virgin to the Scales, as shown in the Star-Zodiac diagram (kindly provided by Robert Powell).[299] The Sun is here moving counter-clockwise.

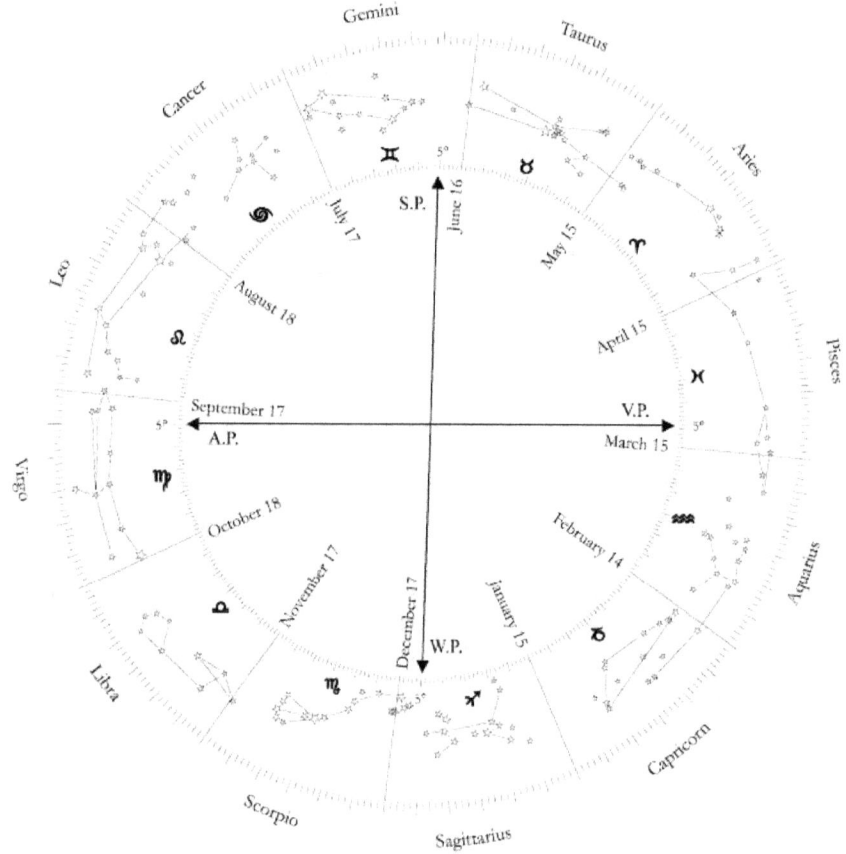

Fig 8.1   Solstices and equinoxes in the star-zodiac.

---

299 One is startled to notice that Powell's famous image of the star-Zodiac (Chapter 7) has here been adjusted, with the Virgo constellation shortened so that it remains within its sidereal 'sign' boundary, i.e it ends at the star Spica. Enquiring, Robert told me that his colleague the late Peter Treadgold was responsible for the adjustment: 'The redesign is based on Egyptian depictions (Neugebauer-Parker, Egyptian Astronomical Texts, volume III - Plates) showing the Virgin to be an upright standing figure and the constellation of Libra having also an upright standing figure holding the scales.'

The bright star Spica is on the boundary between these two signs, indeed it might define this boundary. This is traditionally the most fortunate of stars and connected with the notion of abundance. 'Ceres' (aka the Virgin) is holding it, as a sheaf of corn. Wine-growers might well benefit from harvesting on this day, either 17th or 18th of October, for Sun-conjunct-Spica. The moment when the Sun meets a star moves around by a day or so, and that's because insertion of the leap-year every four years causes such a position in the Sun's yearly journey to move a day or so, year by year, just as the Vernal Equinox moves around the date of March 21st.

Bearing in mind the traditional, solar rulership of the vine, we inspect the Sun's yearly journey: it enters the spring sign of Aries the ram on April 15th, and leaves it on May 15th (within a day or so), then enters Leo on August 18th and leaves it on September 17th. Any planetary or lunar trines to the Sun in these months would enhance the effect of the warmth-trigon. The arrows in this diagram show the Sun's position at equinoxes (horizontal lines) and solstices (vertical lines).

Under azure skies in the Loire valley, there are two wines now produced using a lunar calendar: Joly's and Noel Pinguit's Le Haut Lieu. This last has won every award going (described as 'a stunningly intense, joyful wine' - solar qualities, perhaps?). Over 10% of France's certified organic vineyard area is now Biodynamic[300] (and that amounts to some 1% of total French wine). Joly expresses his views in rather solar terms:

> When we look at a flower or fruit, it becomes perfectly clear that they owe their beauty, their color, their fragrance, their variety of shapes and flavors to the Sun. And it is precisely this power of expression, which manifests itself in constantly new variations that must again be allowed to flourish again in wine - and in every foodstuff.

Readers tired of Euro-plonk will surely appreciate this comment:

> Our apparently progressive agriculture has largely destroyed the soil as a living entity. As a result, the soil is now hardly capable of sustaining growth. It has consequently become dependent on chemical fertilizers, which are inevitably absorbed into the vine itself... In the past, wine growers enriched the soil whereas nowadays, they feed the vines directly. This amply explains the ever-increasing uniformity of wines available from the retailer. As the grapes ripen a critical situation develops in the last few weeks, crucial for final quality. The grapes have to be collected at their optimal stage of ripeness. As acidity gradually decreases the sugar content rises. After harvesting they are crushed and the mix

---

300 Monty Waldin, *Biodynamic Wines.* 2004, p.111.

poured into barrels. That moment needs careful choosing, being the 'birth-moment' of what will mature into wine, so make sure it's a fruit-day.

Maria Thun has played a key role in the development of French Biodynamic wine, with her visits to Joly in the South of France. When I went to visit the Thuns, her research plot of vines was being treated with various sprays of different copper concentrations. Copper sulphate is regularly sprayed onto vines as an insecticide, but she was using higher dilutions of copper, in other words much weaker concentrations along the lines of homeopathic medicines, to investigate their efficacy in the battle against infestation.[301]

One hears it said that the wine drunk by Frenchmen helps them to avoid heart-attacks. They have a far lower incidence of this terminal condition than do the English. Biochemists scrutinize French wines to see which ingredient could exert this health-bestowing effect. It seems to be red wine rather than white which has this reputation. Could this be a meaning of the solar rulership of wine, in view of the Sun's traditional connection to the heart?

Picking Time

That delicious beverage 'Norfolk Punch' avers that its herbs and spices are picked by lunar phases, which might affect its medicinal virtues. I asked Weleda medicines whether they have any policy on when their herbs are to be picked in this respect and could obtain no reply: I believe that they do not. As Kolisko remarked in discussing the comments on this matter to be found in Culpeper's herbal, it is a vanished art.[302] Culpeper's Herbal asserted boldly that 'Physic without astronomy is as a lamp without oil'[303] [NB, 'Physic' then meant, medicine], however one doubts whether the modern-day 'Culpeper' herbal shops take cognizance of this vital principle in preparing their medicines.

Culpeper's Herbal naturally did not comment upon planting herbs, as they grew wild, but solely on when to pick them. He advised:

> At the time of collecting, let the planet that governs the herb be angular, and the stronger the better. If possible in herbs of Saturn, let Saturn be in the ascendent. In collecting herbs of Mars let Mars be in mid-heaven, and let the Moon apply to it in good aspect...[304]

There are three components in this instruction, one of which we have already come across in Dariotus. Firstly, that the Moon should be in aspect

---

301 This section, from 'Under azure skies' is lifted from my 1999 lunar gardening manual.
302 E and L. Kolisko, *Agriculture of Tomorrow*, 1978, p.51.
303 *Culpeper's Herbal,*, ref.10, p.416.
304 Ibid, p.401.

to the ruling planet, secondly that the aspect should be applying and not separating, and thirdly that the ruling planet should be 'angular.' By 'angular,' Culpeper meant, near to one of the four major points in the daily cycle (the horizon, or the vertical M.C. - I.C. axis, i.e. the highest and lowest points reached in the sky). The Moon should be 'applying' to the relevant planet, which pertains to the belief that an aspect is strongest when it is forming, as the two spheres approach their moment of exactitude, and not when they are 'separating,' i.e. moving out of that angular relationship.

A regular gardening column featured in 'The Astrologer,' as was published at the end of the 19th century. This regularly specified the two hours or so of moonrise as the optimal time for gardening operations.[305] For the picking of herbs, it gave times for which, I found, the ruling planet was rising or culminating and harmoniously aspected, but 'free from the beams or body of Mars.' It was thus a continuation of the British tradition given expression in the 'Culpepper Herbal'. The great French herbalist Maurice Messengé said that he picked his herbs at New Moon,[306] and as he always dried them this merely echoes the advice of Pliny. Picking time does become important for the farmer if the seeds are to be collected for next year's sowing.

The work of Agnes Fyfe at the Lukas cancer clinic at Arlesheim, near Basel, involved studying the life-energies of various herbs such as mistletoe and hellebore, in relation to their medicinal virtues. She used chromatographic methods with salts of gold and silver. Her work has been criticized as lacking an adequate quantitative basis, which made her results rather difficult to communicate to others, yet it remains a significant pioneering effort. How did cosmic events, such as solar and lunar eclipses, effect plant sap 'vitality'? She used the pictures which saps made on filter-papers to interpret these things. I once visited her laboratory, where temperature and humidity were carefully controlled, and was rather envious. Concerning the time of picking, she concluded:

> Our experience has shown that on loss of contact with the soil of the Earth, the otherwise continually changing dynamic impulses in the sap are brought to a standstill. The test picture shows the state they are in at the moment of gathering.[307]

---

305 'The Astrologer, composed of twelve monthly parts treating on the Science of Astrology, Medical Botany etc.', was a Victorian journal edited by P. Powley, 'Professor of Astrology and Medical Botany.' Its monthly section 'Advice to Gardeners' selected the Moon rising, in a water sign and waxing.
306 Maurice Messengué, *Of Men and Plants, The autobiography of the world's most famous plant healer*, 1972, New York, p.9.
307 Lehman, 1992 p.13. Lehman points out that the Medieval Arabic astrologer Al-Biruni viewed the rose as Mars-ruled, so traditional sources were not entirely unanimous on this matter. For discussion of the rose-pattern formed by Venus in sidereal space every eight

This supports in a general way the notion that the time when a herb is picked affects its medicinal function. It is not easy to draw more definite conclusions from Fyfe's work.[308]

The Vine, the Rose and the Apple

Let us return to the 'Farmer's Moon' program, which will come to be used by farmers in futurity. If ornamental plants are to be sown, the computer first selects the 'Air' element Moon-sign periods - the stars of Libra, Aquarius and Gemini - over the given sowing interval.

Within these it seeks out any Moon-Venus aspects, and then for that day it would seek out the times of rising of Venus and/or the Moon. The word 'horoscope' means the 'hour of rising,' i.e. that which is rising above the horizon. If on the other hand the plant was one that became woody, i.e. a perennial, the program might place less emphasis upon the sidereal cycle, and instead seek out optimal Saturn aspects. A series of choices are here involved, selecting in turn the week, the day and then the hour.

Perennials such as the rose, the vine and the apple generated a high degree of accord amongst the ancients over their rulership - as indeed did the onion, 'ruled' by Mars, found by Dr Lee Lehman's study of traditional rulerships.[309] A gardener having some apple trees to plant would consult the 'Farmer's Moon' program to find an optimal date. One does not launch a ship at low tide, and who would be so foolish as to plant an apple tree without first selecting an appropriate time? Feasible planting periods are selected over, say, a couple of months, specifying any holiday dates etc not available.

Apple trees come under the rulership of Venus, and the mythology, pentagon geometry and and botanic proximity of apple trees to the rose

---

years, which can also appear as a pentagram pattern woven in the sky, in the context of its 'rulership' of the rose (N.K. 1996, p.153), see the author's 'Venus the Path of Beauty' at newalchemypress.com, or online 'Venus the Rose and the Heart'.

308 A summary of Fyfe's work by Spiess alluded to the course of the year, eclipses, and the synodic and anomalistic (apogee-perigee) monthly cycles: Spiess 1994 'Band 3,' p.33. See especially her 'The signature of the planet mercury in plants, capillary dynamic studies,' *The British Homeopathic Journal*, 1974, LXIII, pp.26-60, 28 for how plant formative forces respond to celestial events, remembering them at the time of picking. On a visit to the Hiscia Klinik at Arlesheim (which develops a cancer-treating medicine using mistletoe), I was told it uses Agnes Fyfe's advice to avoid Uranus aspects for picking times (1984, no English translation).

309 Lee Lehman, *The Book of Rulerships* Whitford US, 1993. Dr Lehman found that traditional sources had little unanimity over Mercury-rulerships: there is an interesting comparison here with the in - depth study of Mercury by Fyfe in the previous reference.

family enables us to credit this without too much difficulty.[310] An apple tree requires periods when Venus is well-aspected to begin life - for grafting or germinating - the more strongly aspected the better, perhaps to Saturn and/or Jupiter. Venus at culmination may be relevant in this context, when it is shining brightly as the Evening Star.

Aspects thus scored would define periods lasting a week or so, including sextile aspects as Dariotus recommended, as otherwise one might be short of the required planetary events. Within these periods one would seek out the lunar aspects, which come and go within a day.

Some readers might like firm square aspects despite the above comments, and indeed there is little experimental evidence to guide us here, merely tradition plus experience. One would like to see an apple orchard planted with half of the trees put into the ground at such a 'right' time, while others were put in at a date when Venus was afflicted, to observe any difference between them as they grew up. Then one might have a quite visible indication of cosmic influences. One should be able to taste the difference!

For roses, a trine between the Moon and Venus may be suitable, as occurs twice a month, this aspect being associated with harmony. To help win prizes, an hour of day could also be chosen, the hour of Venus rising for planting a bought rose, or for grafting the stock required onto the wild rose root. In summary, one has a period of six weeks or so in the Spring for planting, or in the Autumn for grafting. A suitable planetary aspect may give a week or so as a window of opportunity, within which lunar aspects may come and go in a day, where the Moon forms a major aspect to the ruling planet. Lastly, the rising of the Moon or ruling planet gives the optimal hour of day.

Let us conclude with some wise words from the ever-popular seventeenth-century herbalist Nicholas Culpeper, concerning the harvesting of seeds:

> Let them be full ripe when they are gathered, and forget not the celestial harmony before mentioned; for I have found by experience that their virtues are twice as great at such times as others: 'There is an appointed time for every thing under the sun'.[311]

Tree Bud Rhythms

The trees are the Earth's unceasing effort to speak to the listening skies. - Rabindrath Tagore

---

310 However, Thun gave apple trees as 'having a connection with the Jupiter forces' as regards their sowing time. (*Work on the Land*, 1991, p.57). With the exception of Kolisko, anthroposophists have to an alarming degree ignored traditional rulerships.
311 Culpeper op.cit., p.402.

During the winter months, tree buds silently shift their shape to a tidal or fortnightly rhythm. This has emerged from work carried out since 1982 by the mathematician Lawrence Edwards. From his earlier studies of plant morphology Edwards was led to a notion of what he calls λ (lambda) as a coefficient of bud form, where a higher value indicates a more pointed shape[312].

This function decreases as the bud opens in blossom. His book The Field of Form[313] gives values for this coefficient measured daily over several months. He normally finds that this function has what one might call a tidal i.e. fortnightly period of approximately 14 days (tides are twice-daily, but these peak at the 'spring' tides once a fortnight). He and his colleagues have taken many thousands of such measurements over the years.

His theory is that these periods are not quite synodic, but are linked to the planet traditionally supposed to be ruling the tree, e.g. Saturn for the beech. Thus his book presents these fortnightly rhythms as evidence for the concept of planetary rulership of trees. It seemed to me that a computer wave-analysis ought to be able to cast some light on this issue, chiefly by estimating the period of any waveform present. Edwards kindly provided raw data for this purpose, and it was evident that no computer processes had been applied, e.g. the moving averages had been computed by hand.

How the λ-values are obtained need not here concern us. Edwards and his several co-workers obtained a substantial grant for equipment whereby the buds could be photographed under standard conditions every day, and from several such photographs this parameter is derived, each day. It is worth noting that his early books were solely interested in the use of geometrical curves as depicting plant morphology, and his discovery that some meaningful fluctuation was going on within this was more recent.[314]

For four months or so during the winter, the buds are fully formed and not doing anything much, in a more or less static condition. This is the period when Edwards has noticed this fluctuation in shape. For some years I took

---

312 For a fine introduction to this topic, see John Blackwood, 'Geometry in Nature, Exploring the Morphology of the Natural World through Projective Geometry', 2012, Chapter 13.
313 Lawrence Edwards, *The Field of Form*, 1982. For discussion of the idea of 'path-curves' from which the λ-function is derived, see *Projective Geometry -- Creative Polarities in Space and Time*, Olive Whicher, 1971, page 158 et seq. For experiencing path curves one may try the exercises in Edwards's book *Projective Geometry,* Floris, 2003. Also algebraic treatments of various categories of path curves are given in the lengthy appendices of Edwards's *The Vortex of Life*, Floris, 2006 (thanks to David Heaf for this advice).
314 In *The Vortex of Life*, 1993, Lawrence Edwards concludes that the fortnightly lambda-rhythms of the tree-buds do not remain in phase with the lunar conjunctions and oppositions to their supposed planetary rulers, but drift away from these - a 'phase shift,' attributed to a backward-streaming flow of 'negative' time (p.261). I don't recommend this argument.

no interest, as it was a subject of no practical use, and the bent of my work has been towards matters of practical use to gardeners. However, Edwards was the only person in the English-speaking world doing plant-Moon research in the 1990s, so I wrote to him.

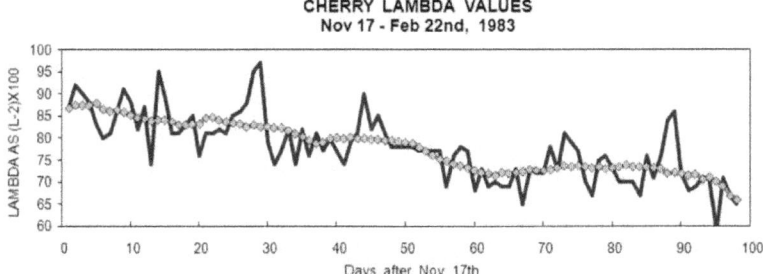

Fig 8.2 The 'lambda values' of buds on a cherry tree, measured by Lawrence Edwards over four months in the winter of 1982/3.

Let us take two trees whose rulerships would seem to be fairly straightforward: the beech (Saturn) and the cherry (Venus) - although Rudolf Steiner assigned to the latter a lunar rulership.

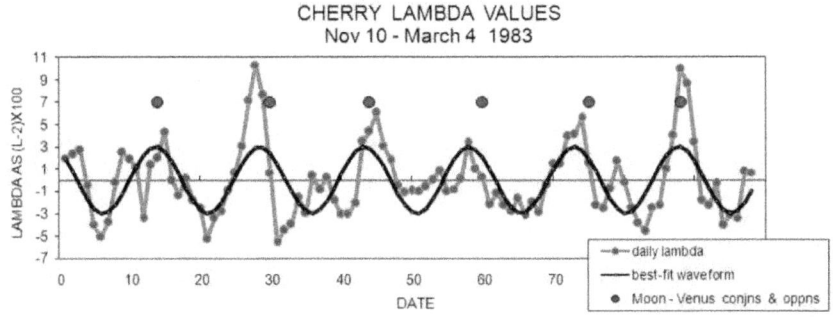

Fig 8.3 Likewise cherry lambda values, with best-fit sinewave, and fortnightly Moon-Venus conjunctions and oppositions.

Over several months, the mean period of the 'Saturn-month,' ie between successive periods of Moon-conjunct Saturn, is 27.4 days.[315] During the winter of 1984/5, Edwards took λ-values from two adjacent beech trees, which were co-varying in a fairly similar manner. I averaged the two

---

315 The equation, $1/27.32 - 1/10759 = 1/27.39$ days finds the mean Saturn-month period, twenty-nine and a half being its orbit-period in years. Its period over the two months of this trial was very slightly shorter, 27.35, as Saturn was initially moving retrograde.

together, and then put a best-fit waveform through them, see graph. Its optimal period turned out to be 27.0 days, or rather half of that. Now, that is slightly shorter than it should be on this theory – it ought to be 27.3 or 27.4 days if it is to remain linked to the Moon-Saturn conjunctions. One can see in the diagram that the peaks are falling a few days before the Saturn-Moon conjunctions and oppositions, also shown in the diagram.

For Venus however that interval between Moon-conjunctions will be 29.5 days: as Venus revolves fairly closely around the Sun, the mean period of lunar conjunctions is identical with the synodic period. We should easily be able to distinguish these periods. For the cherry tree buds, the optimal period over the two months shown was 29.6 days. The diagrams show the cherry data set. That is a very close fit. Summarizing:

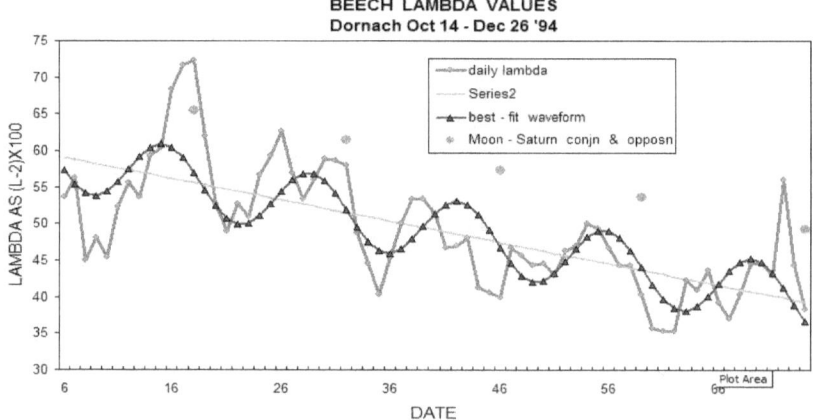

Fig 8.4   Beech tree lambda values, over the winter of 1993/4, with best-fit sinewave.

Cherry buds Nov. 10 - March 4 1984, mean of the daily lambda values 2.78, amplitude of tidal oscillation 1.25%, period half of 29.6 days; in phase with Moon-Venus conjunctions and oppositions (shown on graph).

Dornach Beech Oct. 14 – Dec. 26 1994, mean of daily lambda values 2.49, amplitude of tidal oscillation 1.8%, its period half of 27.0 days.

Dornach Hornbeam II Oct. 19 – Dec. 19 1994, lambda mean 2.4, amplitude of tidal oscillation 2.45 %, its period half of 27.0 days.

The cherry tree data is truly remarkable in that we have seven whole cycles, that is over three months, without a break. This enables us to estimate its period length with great confidence. Its rhythm is clearly tidal, i.e. half of the synodic cycle. We are also impressed by the way its peaks stay well aligned to the lunar conjunctions and oppositions with Venus once a fortnight as shown by the graph. In general however most of the Lawrence

Edwards data-sets extend over only two months, as the observers take a week off over Christmas. Clearly there is a limit to what people can do on a voluntary basis, however this does knock a hole in any technique of waveform estimation. Normally the data contains some fairly strong seasonal trend, and it is necessary to remove this or allow for it in some way before the fortnightly sine-wave can be fitted. There is no legitimate means of putting such a trend-line through data containing a one-week gap!

For the Lawrence Edwards tree-bud rhythm data of Chapter 8 where the buds were found to be altering their shape to a fortnightly rhythm, the period was shown to be slightly shorter than half of the sidereal or synodic monthly periods. Let's write the equations to a first approximation as:

Cherry bud rhythm: $\lambda/\lambda_m = 1 + 0.013\sin2(360 \times T/29.5 - p_1)$ (Fig 8.3)
Beech bud rhythm: $\lambda/\lambda_m = 1 + 0.018\sin2(360 \times T/27.3 - p_2)$ (Fig 8.4)

where T is time measured in days and p is a constant signifying phase – his waveforms are often not very evidently in phase with the designated Moon-planet conjunctions.

Lambda is his key term related to the egg or cones shape and how pointed it is – it's hard to explain or visualize and let's not do that. Traditionally, the cherry tree was 'ruled' by Venus, and because Venus weaves back and forth around the Sun in the Zodiac, crossing it five times in four years, a Moon-Venus period is going to look rather similar to a Moon-Sun period. In this argument (if you want to accept it), Moon conjunctions with the Sun and with the inner planets Mercury and Venus are going to give rather similar periods. Lambda for the cherry tree buds has a mean value of 2.8, and the amplitude term we've found will make this lambda-function concerned with bud morphology fluctuate by 1.5%. Whereas in contrast, the beech-tree period was shorter, and had some value close to half of 27.3 days. The meeting of Saturn and the Moon will have a period comparable to this.

Allow yourself some time mulling over the sheer incredulity of this effect. All the little buds, on all the trees, when you thought they were just surviving the bitter winter winds, are they actually shape-adjusting in tune to finely-differentiated lunar rhythms? Do you really want to believe such a thing? Can a rather small oscillation in bud shape be going on in all buds around the world, over the quiet winter months?

Lawrence Edwards wrote a sequence of three books: the first was purely about projective geometry, written by a school maths teacher, and introduced the lambda concept, which was not his invention. This book had no inkling of what he was later to become famous for, or at least moderately renowned amongst the small group of people who notice such things. In his second opus he had apprehended the fluctuation of this function with the lunar-tidal period, and then his third and final book focused upon how this lambda-function varied in different trees, bushes and flowers to this rhythm. As to whether his work has any practical value, it could well be so, if perchance it were to validate the notion of 'rulership', which in this case would mean, for instance, that cherries were more tasty if the tree has been planted or grafted with Venus's position in mind. We are here suggesting that the research initiated by the late Lawrence Edwards may not yet have got that far.

The net alteration in the λ-value given by this fortnightly rhythm is small, a mere 1-2%, yet his procedure seems good enough for detecting this fairly reliably. As an anthroposophist he has rather odd notions about rulerships, for example that the cherry tree is ruled by the Moon, where tradition gives it a Venus-rulership. However, in correspondence he appeared quite open-minded about these issues. Also of interest is his view that major planetary conjunctions affecting the ruling planet can be discerned from his data, as tending to block out the otherwise regular rhythm. This would indeed be remarkable if it could be established. Here again, a four-month period of data without a week's interlude would be a help.

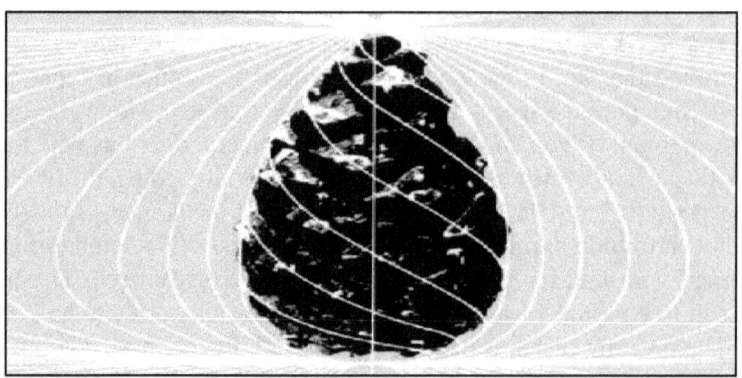

Fig 8.5 Varying lambda-value curves around a fir cone.

Other trees do not fit their predicted periods as well as the two cases shown, and so Edwards' case is by no means established. In general, his trees seem to have periods nearer to the sidereal month than the synodic one. He has clearly shown a tidal, fortnightly rhythm going on in most tree buds,[316],[317] but questions remain concerning his planetary-rulership hypothesis.

---

316 Olga Holbeck, 'Variations in the forms of Plant Buds,' Star and Furrow, Winter 1987 pp.17-20.
317 For comparison, a fortnightly rhythm exists in tree electrical function (Markson 1972; Burr 1977) as it does likewise in patterns of bee activity (M. Oehmke, 'Lunar periodicity in the flight activity of honey bees'1973), see Ch. 4.

As to why Mr Edwards troubled to take all these measurements over the winter months, he explained in a letter:

> The work affects our whole view of our solar system: - is it a mass of separate particles jostling randomwise in space, or is it an ordered and living organism, every part of which acts in living relationship with the whole?

This is indeed the question. I had a prejudice against accepting his conclusions, on the grounds that planets should not exert such a physical effect, and that they should rather influence more the quality or virtue of the plant in some medicinal sense. Time will, no doubt, tell.

9. Horse Breeding & the Lunar Month

[Farmers] notice the aspects of the Moon, when at full, in order to direct the copulation of their herds and flocks, and the setting of plants or sowing of seeds: and there is not an individual who considers these general precautions as impossible or unprofitable. - Ptolemy, Tetrabiblos, Ch. III

In the Middle Ages, the Arabs were the finest horse-breeders in Europe. Their subtle understanding of astrology could possibly have helped them here. In the eighteenth-century, Britain out-bred them, producing the finest racehorses in the world. The UK is presently slowly losing this edge, partly because, with the price of top stallions at several million pounds apiece, they are being brought up by Japanese and American buyers. Hiring a 'stud,' that is a stallion that has performed well in races, is expensive, and if there were an optimal time for doing this, it would be worth knowing about.

Racehorses are the one animal whose sex life is rigorously documented and published. Horse breeding is something which is taken very seriously, so it seemed to me that results obtained in this domain would be of considerable interest. Each year, the 'Statistical Record' is published for thoroughbreds (i.e. racehorses), of offspring between each mare and stallion. I was kindly given permission (via a friend Camilla Power who was then something of a horse-breeding journalist), from an eminent thoroughbred stud-farm, to access the data from their yearly stud-books.[1] Is the fertility of mares, like all living things, conditioned by the changing Sun-Moon angle, or does the inner clock of their three-week estrus cycle, plus an intensive hormone regime, rule this out?

Each mare is 'covered' several times per season, i.e. put together with a stallion, until it conceives, except for the ten percent or so that remain barren. Most of the mares live on the stud-farm for some months, but some are just walk-ins, i.e. brought in to be covered. A modern stud-farm keeps going seven days a week without a break, right through the breeding season. The mares have their mating-season artificially skewed for a purely bureaucratic reason: they start conceiving in February, so that most conceptions take place in March-May, far earlier than would naturally be the case, owing to regulations about the age at which they may start to race. Left to themselves, they would breed over the summer months, because conception takes eleven months and no horses would choose to drop their

---

1 Anon. (1987-2000) *The Statistical Record: Return of Mares*, Northants, Weatherby's.

foal when frost is still around. Modern mares thus receive intense hormonal stimuli to kick-start their estrus cycle earlier than it should, as well as artificial illumination at night-time to make them think it was summer. Thus the gentle effect of full-moonlight could hardly be operative, could it? Let us hope for a future generation of caring horse-race enthusiasts, who insist that no artificial hormones or artificial-light regimes be used for horse rearing.

The compressed breeding season for thoroughbred mares can easily undermine the detection of a lunar-month effect. A smooth breeding-season extending over months would be greatly preferable. The only solution is to keep adding successive years of data until the inequalities gradually average out. A further problem arises with the Moon moving at differing speeds due to its apogee-perigee cycle, so that lunar inequalities will tend to balance out over nine-year periods (that's half of the node-cycle as well as the nine-year rotation of the apse line). For this reason, I suggest that one needs at least nine consecutive years of data in order to have confidence in the results.

The Timing of Estrus

Like cows, horses ovulate to a twenty-one day cycle throughout the breeding season (as do pigs and goats, while sheep come in a few days shorter at 17 days). In the 1980s, vets became able to discern within a day or so when estrus was occurring: mare coverings are now done very efficiently only when the mare is ovulating.[2] It may surprise some readers that horse copulation ('coverings') still takes place, while cows have to put up with just being injected, and perhaps this is an expression of the high esteem in which racehorses are held. I found that in some years the mares would become synchronized in their ovulation, so that the stud-farm became more busy at three-weekly intervals. Figure 1 compares results from two stud-farms, plotting the total number of coverings per four days (the thin line is a best-fit sine wave of 20.5-day period). In this synchrony the mares resemble many creatures in the wild, where the females of a species tend to ovulate in synchrony.[3]

---

2 Peter Rossdale, 1993: *The Horse from Conception to Maturity*, London, Allen, p.14.
3 Martha K. McClintock, 1984, 'Estrous synchrony: modulation of ovarian cycle length by female pheromones' *Physiological Behavior*, 32, pp.701-705.

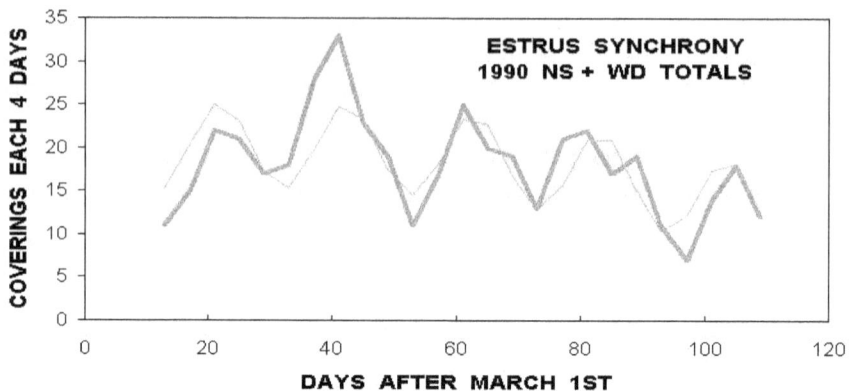

Fig 9.1 Mare covering frequencies, grouped at 4-day intervals, showing the 21-day estrus period, and how mares from two different studfarms at Newmarket both displayed the same trend.

Like cows, horses ovulate to a twenty-one day cycle throughout the breeding season (as do pigs and goats, while sheep come in a few days shorter at 17 days). In the 1980s, vets became able to discern within a day or so when estrus was occurring: mare coverings are now done very efficiently only when the mare is ovulating.[4] It may surprise some readers that horse copulation ('coverings') still takes place, while cows have to put up with just being injected, and perhaps this is an expression of the high esteem in which racehorses are held. I found that in some years the mares would become synchronized in their ovulation, so that the stud-farm became more busy at three-weekly intervals. Figure 1 compares results from two stud-farms, plotting the total number of coverings per four days (the thin line is a best-fit sine wave of 20.5-day period). In this synchrony the mares resemble many creatures in the wild, where the females of a species tend to ovulate in synchrony.[5]

As regards analyzing this data, how does one represent nearly five thousand data points in such a way as to depict what is happening in the course of a lunar month? No method is perfect. The worst method is merely to group the data into four sectors, four lunar 'quarters,' which is often done by those who publish 'null results' in this area. A better method is that of lunar-day numbers, which works as follows. Let us call the first day that begins after the conjunction of the Sun and Moon 'day

---

4 Peter Rossdale, 1993: *The Horse from Conception to Maturity*, London, Allen, p.14.
5 Martha K. McClintock, 1984, 'Estrous synchrony: modulation of ovarian cycle length by female pheromones' *Physiological Behavior*, 32, pp.701-705.

Horse Breeding & the Lunar Month

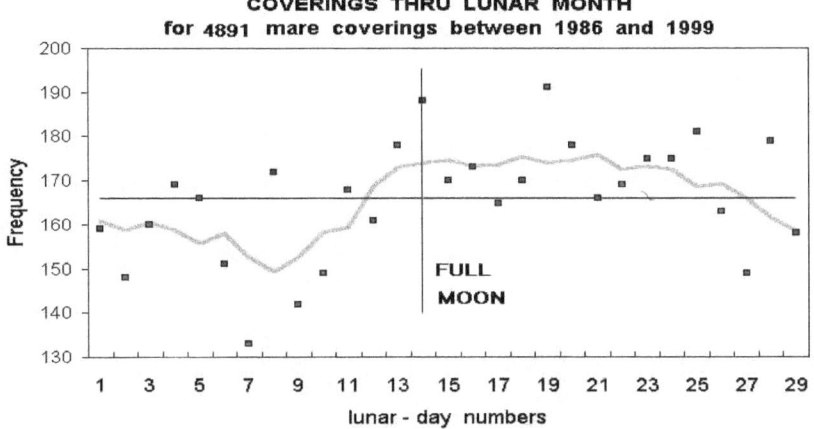

Fig 9.2 Data for mare covering dates over 14 years 1986-99 from one stud farm, plotted by lunar-day number 1-29 at covering date.

one.' We count from that up to day 29. On what day will the Full Moon fall? The answer to that will vary considerably, depending upon the position of perigee, affecting how fast it moves around the sky. So, this method blurs the position of the Full Moon over several days – that is a drawback. The data is grouped according to which of these twenty-nine 'lunar day numbers' it fell into. There is a half-day left over and one discards it, leaving a slight discontinuity over the New-Moon position – or, wherever in the continuous circle one chooses to start the counting.

Figure 2 shows the frequency of covering-dates using this method, putting all 14 years of data together. This indicates mare estrus, when the horses come on heat, because coverings are nowadays done during these days. It shows a distinctive minimum, or a drop in the number of horses on heat, at the first quarter of the waxing moon. This is followed by a plateau of such days, lasting from just prior to the Full Moon to a good week after it. A statistical prediction here would be: an excess of mare coverings lasting at least a week, maybe ten days or so, starting from the day before the Full Moon.

How big is the lunar-monthly trend? To answer that, it is convenient to take five consecutive lunar-day numbers, one set centered on the day after Full Moon, and the other on the day after the First Quarter (i.e. seven days earlier):

Total coverings 1986-99 summed over five lunar-day numbers:

First Quarter 717 / Full Moon 819

Expected 722, Difference = 14%

We are here comparing roughly one-third of the total data. Thus, the frequency of mares coming into heat swings through fourteen percent in a normal lunar month. That is the most significant discovery reported in this volume. It is probably my one original contribution to the topic.

Mare fertility

Flocks brighten the mountains, Herds throng up the valley, Wild beasts fill the forests. - William Blake, Vala, or the Four Zoas

Let 'fertility' for a group of mares signify the proportion of coverings that lead to a registered conception. To measure this we have to compare data from two different sources: the studbooks kept on the stud-farm, in which every covering is recorded, as well as a record of conceptions, but that may not be quite reliable enough - with the official, yearly-published 'Statistical Record', which records the outcome for each mare per season: given as barren, aborted, no-return or as having a live-birth. 'The General Stud Book' also gives this data, more from the point of view of the stallions.[6] One tries to minimize the 'no-return' data, for which no outcome is recorded (usually because the mare went abroad). All of this 'no-return' data, i.e. mating-pairs of no known outcome, has to be removed before an evaluation of mare fertility can be made.

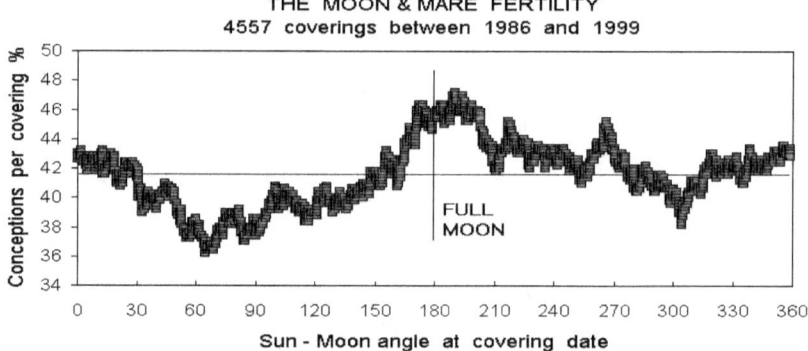

Fig 9.3 The proportion of mare coverings which led to conception, over the same 14 years of data, with a moving average plotted around the lunar month.

Figure 9.3 shows how the fertility of mares varies with the lunar month. The total has slightly decreased because the 'no-return' coverings have been discarded. Overall, the value of this parameter hovers around forty percent – i.e., four out of ten coverings lead to conception. It shows a more distinctive peak around the Full Moon than did the previous graph – and, the pronounced first-quarter minimum remains. The latter monthly dip in

---

6 Anon: 'The General Stud Book,' Vols. 43 (1997), 44 (2001), Northants, Weatherby's.

fertility has never been even hinted at before, in the thousands of years of tradition of horse-breeding.

The graph is showing something like an eight-point swing in mare fertility as a percentage, through the lunar month. If we take the five lunar-day means as used above, then the monthly swing here appears as just over five percent. This graph has no minimum at New Moon, fourteen days away from the Full, so this effect cannot be approximated by a simple waveform. It shows, rather, a period of ten days or so in which the fertility of mares swings from its minimum to a maximum. The graph has been smoothed using a 'moving average'. (A 450-point moving average was used, where each point plotted is at the center of that mean. This moving average is around ten percent of the total; the larger it is, the more smoothed-out the graph appears). Each individual data-point has a Sun-Moon angle for noon of its covering date, and is scored one or zero according to whether conception occurred.

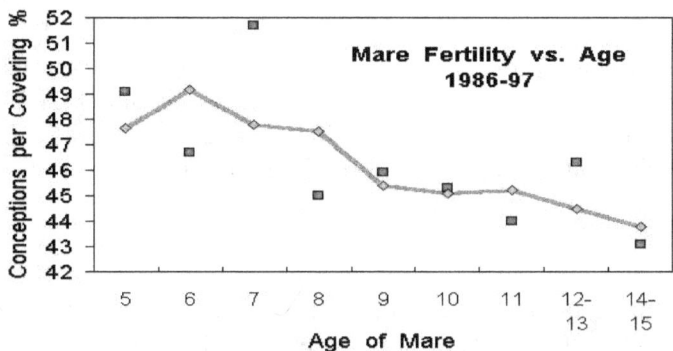

Fig 9.4   Showing how mare fertility peaks at 6-7 years of age.

Results from the first nine years of this data were published as a pilot study in the 'Equine Veterinary Journal'.[7] As far as I could tell, no-one believed it, or at least no-one took any notice.

The director of Newmarket's Equine Fertility Unit required that the data be separated by age. The fertility of mares under twelve years of age was plotted, as compared with the older ones (showing no very noticeable difference in regard to their lunar-phase curves) and that methodology has here been continued. A couple of follow-up articles have been published.[8]

How mare fertility varies with age is shown in Figure 4 (smoothed using a three-point moving average). Articles on this subject describe mare fertility

---

7 N. K and Camilla Power (2000): 'The influence of the lunar cycle on fertility on two thoroughbred stud-farms,' Equine Veterinary Journal, 32, pp.75-77.
8 N.K., 'A Lunar Cycle in Mare Fertility', Correlation, 2011, 27 (2), pp.38-47. See also ref. (131)

as decreasing linearly with age and seem not to allude to the intriguing peak in fertility here evident around 6-7 years. In other words mares are not like cars where the value starts to drop as soon as they are used, but rather their fertility takes a year or two before it reaches its peak.

Geomagnetism and Sun-node angle

Twice-yearly, the 'eclipse seasons' turn up, when solar and lunar eclipses take place. The Sun in its journey round the zodiac then crosses over the node-axis, the axis linking the two positions where the Moon crosses the ecliptic each month. The Sun moves round the zodiac once a year, and the lunar-node axis revolves much more slowly, once per 18.6 years, in the other direction.

Traditionally the nodes were regarded as powerful energy-points in the Moon's cycle ('Dragon's head' and 'Dragon's tail') and large-scale American investigations have confirmed this[9],[10],[11]. There is evidence that this node cycle affects long-term trends in climate[12] and wheat production.[13]

Simon Best has for long argued that the nodes modulated the effect of the syzygy via their influence upon the Earth's geomagnetic field.[14] His work has been mainly in human psychology as well as co-authoring a lunar-planting manual with me in the early 1980s. We all know that the Earth's magnetic field points towards the North Pole, but in addition it fluctuates greatly in magnitude from day to day; it's a very mutable thing. Its strength tends to peak over the Full Moon, or on the days just after it.

It responds maximally to the Full Moon when syzygy falls near to the node axis, during the months of the 'eclipse seasons' (the position of 'syzygy,' the axis on which the Full and New Moons take place, is indicated by the Sun's position in the zodiac). Researchers at the 'Massachusetts Institute of Technology' (M.I.T.) found this out from analyzing several decades of geomagnetism data.[15] Best argued that 'strong Moons' as regards stressful

---

9 H. Stolov and A. Cameron (1964), 'Variation of Geomagnetic Activity with Lunar Phase,' *J Geophys Res.*, 69, pp.4975-4982 (The Goddard Institute for Space Studies).
10 B. Bell and R. Defouw (1964): 'Concerning a Lunar Modulation of geomagnetic Activity,' *J Geophys Res.*, 69, pp.3169-3174 (Harvard College Observatory).
11 . O. Schneider (1967): 'Interactions of the Moon with the Earth's Magnetosphere,' *Space Science Reviews*, 6, pp.680-704, 682.
12 Robert Currie (1987): 'Examples & Implications of 18.6- and 11- yr Terms in World Weather Records,' In: Rampino M, Sanders J, Newman W, Konigsson LK, eds., *Climate: History, Periodicity and Predictability*, New York, Van Nostrand.
13 Robert Currie and S. Hameed (1986): 'Climatically Induced Cyclic Variations in United States corn Yield and Possible Economic Implications,' *Cycles* (Pittsburgh), May/June, pp.78-84.
14 Cyril Smith and Simon Best (1989), '*Electromagnetic Man*, Health and Hazard in the electrical Environment,' New York, St Martin's Press, pp.41-2.
15 B. Bell and R. Defouw (1966): *Jnl. of Geophysical Research*, 71, 'Dependence of the lunar modulation of geomagnetic activity on the Celestial Latitude of the Moon,' pp.951-

human experience were those Full Moons near to the nodes, (of low celestial latitude), i.e those associated with maximal geomagnetic perturbation.

In Chapter Two we looked at three graphs (Fig 2.4) showing how geomagnetic activity peaked at Full Moons: it varied crucially with the celestial latitude of each Full Moon (as found by Bell and Defouw, 1966). At zero latitude, the top graph, the variation was greatest. The bottom graph shows the least activity at maximum latitude, i.e. with zyzygy furthest from the nodes.

For years, I had surmised that some kind of node-key must exist, although it would only become evident through evaluating years of data, owing to the slow motion of the lunar nodes. When finally I acquired enough data to be able to do the analysis (which took a long, long time), a result appeared, the opposite of what I had been expecting.

The modulation was the reverse of that seen in geomagnetism, where the greatest effect of the lunar month takes place far from the nodes, i.e. with syzygy at high lunar latitudes. In other words, the greatest lunar-month effects in the horse data was found when the Sun was furthest from the lunar nodes. To do this analysis I first removed mares twelve years or over; then, as the figure shows, the swing in fertility appears as at least ten percent. Horse-breeders take for granted that age is the biggest factor in determining mare fertility, however it can be seen that the amplitude of this 'strong-Moon' effect appears as even larger than that due to age. It's around ten percent, which is huge.

I urge readers to view this graph with some degree of incredulity: can such a huge modulation in mare fertility really exist, that is astronomically determined, and unknown over aeons of time? Does it apply to us humans too? I went on collecting the data over 14 years because I suspected that there must be some long-term pattern in the lunar-month response, and this is it.

Summarizing, let's go through the stages in processing the data: 'no-return' cases were removed and also mares over a certain age were also excluded, being less fertile. For each covering date one finds the Sun, Moon and lunar node longitudes for noon of the date in question, and thence one ascertains the lunar-day number; one sorts the data by Sun-node angle, dividing it into two groups, according to how far the Sun was from the lunar nodes.

Then, each group is sorted by Sun-Moon angle, and plotted as graphs. That half of the data at low celestial latitude showed hardly any modulation by the lunar month, except for a dip in the first quarter. A paper on this has

---

967; 'On the Lunar Modulation of Geomagnetic Activity,' pp.4599-4603; 'Discussion of Paper ... The Lunar Period, the Solar Period and Kp,' pp.5770-5773.

been published in the 'Journal of Biological Rhythms', formerly the Journal of Interdisciplinary Cycles Research.[16]

To be sure, the data can be analyzed in various ways. We might want to select only the first covering per mating pair, which comprise 38% of the total. At a cost of losing a large proportion of one's data, this does give a more homogeneous data-set: the likelihood of conception differs between the first and subsequent coverings and they are distributed differently through the breeding-season.

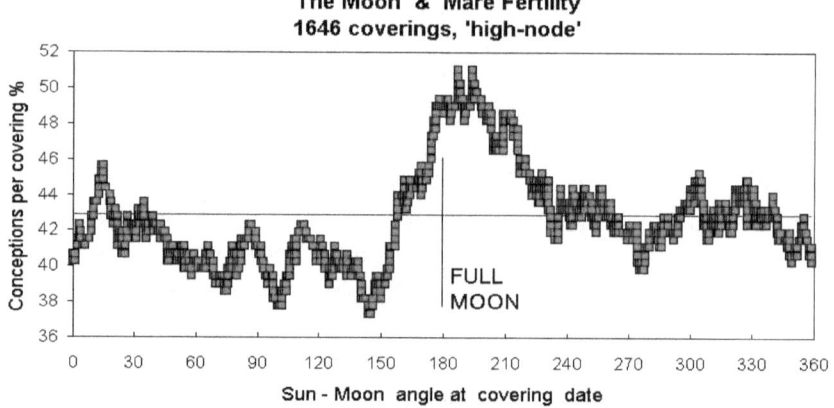

Fig 9.5 As per figure 9.3, but selecting the 'high-node' half of the data (compare Fig 2.4)

The graph is here shown. In addition this graph also shows the frequency of live-births, following eleven months later after the covering date plotted. The percentage is somewhat lower, owing to the small proportion of conceptions that result in abortions or stillbirths.

I never found traditions of mare fertility as varying with the lunar month. The beautiful truth is, that during the mare breeding season, a three-week 'endogenous' rhythm is interacting with a four-week exogenous rhythm, to define the peak fertility experienced by mare populations. Why are these delicate lunar rhythms not wiped out by all the sex-hormones and night-lighting that modern breeding requires? That's easy to answer: Mother Nature is tough. But, there is evidence from other sources that lunar-month cycles in animals and insects are altered by tampering with such ambient conditions.[17] Taking these results at face value, they show a swing in the frequency of mares coming on heat of twenty percent in the lunar month,

---

16 NK, 'Lunar Effect of Thoroughbred Mare Fertility: an analysis of 14 years of Data 1986-1999' Biological Rhythm Research 2004, 35, pp. 317-328.
17 M.G. Oehmke, 'Lunar Periodicity in flight Activity of Honey Bees,' *Jnl. Of Interdisciplinary Cycles Research,* 1973, 4, pp.319-335.

and of conceptions forty percent, between First Quarter and the Full, for the half of the data away from the 'eclipse seasons.'

Table 1: 'Strong Moons': Excluding mares over 12 years of age, then excluding that half of the data with Sun-node angle < 45° (809 mating-pairs, 1647 coverings).

High Node'	Cover	Concn	Ferty
1st Quarter	243	99	40.7%
Full Moon	296	140	47.3%
3rd Quarter	287	119	41.5%
New Moon	254	112	44.1%
Xs FM/NM	17%	25%	
Xs FM / 1st Q.	22%	41%	

Table 2: 'Weak Moons': As above but for Sun-node angle < 45°; (847 mating-pairs, 1781 coverings).

Low Node'	FM/1st Q.	Concn	Ferty
1st Quarter	286	119	41.6%
Full Moon	307	136	44.3%
3rd Quarter	309	137	44.3%
New Moon	302	132	43.7%
Xs FM/NM	1%	3%	
Xs FM/1st Q.	7%	14%	

Do these results apply to all racehorse mares, or to all horses or all mammals, or indeed do they apply to all living animals and to man as well? With human fertility continuing to plummet in 'developed' nations, this question could be of some concern. No-one has any human data on this subject, because there is no human stud-book!

One hears it said that human fertility, as above defined, is around ten percent. One day, I believe that equations will describe these rhythms: the 'Silver Axioms.' To be able to do this, a further set of comparable data, preferably from a different climate – antipodean and over the same 14 years – is probably required. There must exist, I suggest, equations of universal validity governing the alteration of mare fertility with age and the lunar month.

When I tell people about these horse studies they always ask the same question: 'What is the mechanism?' I try hard to understand: a mechanism is something made out of metal, whose parts touch each other. I suggest that the connection of fertility with the Sun-Moon angle is a primary phenomenon, which does not have anything behind it that 'explains' it. When scientists talk about 'explanation' it means that they wish to describe something living in terms of a model in their mind that is non-living, a habit which may not always be helpful.

The public, it seems to me, have a great interest in these matters, and always want to be told about any results in this field; while 'experts' have a rather contrary disposition and seem to feel it is their business to disparage the subject. Is the latter attitude a gender prejudice, arising from male scientists' unease over the lunar-month- type cycle going on within womankind?

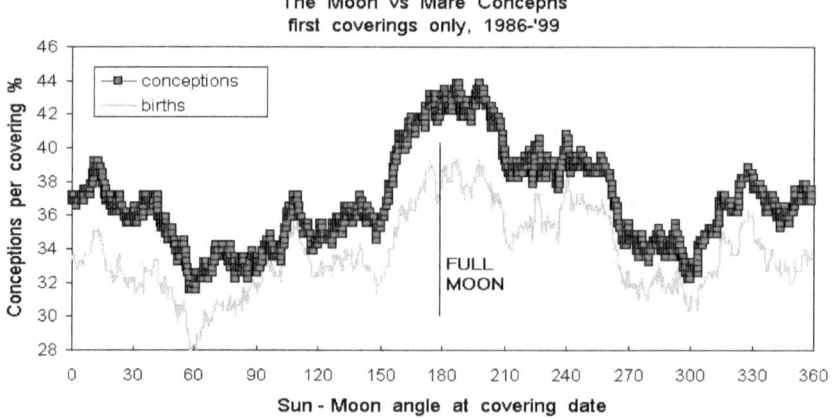

Fig 9.7 as Fig 9.3, but selecting only first coverings of each season per mating-pair; also plotting the % of live births per covering.

Whatever the answer, my hope is that the effect here shown is large enough to be of direct economic value to any horse breeders who wish to apply it.

Champion racehorse stallions are valued at several million pounds, and anything that can increase the likelihood of a successful conception may be seen as worth trying. Also I'd like to hope that Arab horse breeders may not have so strong a prejudice against practical use of the lunar-month cycle.

What is called 'chronobiology' is a rather Satanic science which (I suggest) in practice involves or justifies the ceaseless flow of blood of small furry animals, funded by drug company profits. One can hardly open a journal of chronobiology without a sense of horror. Biologists are conditioned to believe that they are not doing real science unless some organism is cut open. In earlier chapters we quoted the late professor Frank Brown to the effect that lunar-month studies were of especial value here, in showing the need to discern a more exogenous (externally-generated) cause of the rhythms involved.

Indeed chronobiology will only begin to exist properly when scientists compare the effect of solar and lunar rhythms upon the life of an organism. An awareness of the lunar cycles brings romance and mystery into our relation with the horse, which is I believe well merited. The lunar nodes have a huge effect in modulating the fertility-rhythm of the lunar month and vets are going to have to get used to this.

The Spring Surge in Trout-Migration

Our attention now turns to a more aquatic response to lunar cycles, a topic treated with great thoroughness by a work recently translated from the German, 'Moon Rhythms in Nature'.[18] Trout in an aquarium grow to a rhythm whereby their weight peaks just before the Full and New positions.[19] This was shown using several hundred small trout kept in a laboratory, and weighing them every four days as they grew - a commendably simple experiment.

Fish respond biochemically to the lunar cycle: salmon and certain trout species hatch in fresh water, and at a certain stage of their lives transform to become ocean-dwelling salt-water fish. Salmon fisheries need to be able to predict when this event will occur, as they have to release hatchery-reared fish into the river immediately prior to this. Thyroid hormone thyroxine triggers this big moment in the salmon's life, and there is a specific New Moon in the Spring which times this hormone surge.[20]

If the salmon wish to swim down the river without being seen by predators then a New Moon is the optimal time of month for them. This was discovered by zoologists at the University of California, who concluded that

---

18 . K. Endres and W. Schad, *Moon Rhythms in Nature*, Floris 1997.
19 K. Farbridge & J. Leatherland, 'Lunar Periodicity of Growth Cycles in Rainbow Trout,' *Journal of Interdisciplinary Cycles Reserch,* 1987, 18, 169-177.
20 G. Grau et.al, 'Lunar Phasing of the Thyroxine Surge...' Science, 1981, 211, 607-9.

a lunar calendar was essential for efficient culture of this economically valuable resource.[21]

The New Moon at the Vernal Equinox triggers this big moment in the salmon's life, called 'smolting,' and prepares them for downstream migration and eventual survival in saltwater. One group of researchers scheduled the release of salmon from hatcheries at the New Moon, finding that over several years an enhanced survival rate was associated with New Moon releases as compared with other phases of the Moon.

In Florida, there is a common fish the 'Inland Silverside' which has a semilunar rhythm of spawning. Fish were collected in the wild and maintained in a laboratory, where they still showed their fortnightly peaks in spawning synchronized with syzygy (days of Full and New Moons).[22] A different study of the anemone-fish found that it spawned at the lunar quarters, while incubation took seven or eight days, so that hatching peaked at syzygy.[23]

In the 1920s, an American called John Knight developed his theory about when to fish, and his theory has been incorporated into popular almanacs. His theory was that fish feed only twice a day, when the Moon culminates (reaches its highest point) and reaches its nadir (lowest point, below the horizon). Some confirmation of this theory comes from the laboratory study of a small tidal fish which feeds on barnacles off the West coast of Britain. Kept in the laboratory, the fish showed increased swimming activity at the time of high tides, i.e. every 12.5 hours.[24] A weaker, solar rhythm of activity was also present.

Research by Amanda Vincent at Cambridge University biology department concerned sea-horses. This showed that sea horses continued to lay their eggs at Full Moon even when in laboratory tanks, deprived of any evident source of information on the lunar cycle. This was claimed on a television program but never published by Dr Vincent.

I visited the laboratories in question - it strained my credulity to think that Cambridge University would possibly allow research on such a topic - and, although she had by then departed the staff did confirm that, indeed, she had been investigating the subject. Had the TV program assisted her departure, I wondered?

---

21 R.S. Nishioka at al., 'Effect of lunar-phased release on Adult Recovery of Coho Salmon ... in California' *Aquaculture* 82 (1989) pp.355-365.
22 R.M. Ross, 'Reproductive Behavior of the Anemone fish' *Copeia* 1 (1978) pp.103-107
23 M.T. Sherill and D. Middaugh, 'Spawning Periodicity of the Inland Silverside .. in the Laboratory: Relationship to Lunar Cycles' *Copeia*, 2 (1993) pp.522-528.
24 R.N. Gibson, 'The Tidal Rhythm of Activity of Coryphoblennius Galerita,' *Animal Behaviour*, 18 1970, pp.539-543.

The color sense of certain fish have been shown to vary with the lunar month.[25] In 'comprehensive and long-term experiments,'[26] guppy fish were shown to see yellow best at the Full Moon and violet best at the new moon. A fish's world changes color during the lunar month! At the 'hidden' time of the month when Selene's sphere is invisible, the mysterious hues of violet grow stronger in Neptune's realm. The method was described as 'exacting' by Burns.[27] Their results are echoed in human studies showing a shift towards enhanced blue hue perception during the New Moon and the converse at Full.[28]

GMF

Modern astronomers are keenly interested in the Earth's geomagnetic field, but none of them would dream of considering the manner in which this is modulated by the lunar cycle. That would be quite beyond the pale. It is not that there is anything wrong with the several papers here alluded to. However, they were done several decades ago and maybe that is what enables them to be ignored: back in the swinging sixties papers on a lunar theme were OK. My opinion, for what it's worth, is that the public would have a lot more interest in the subject if astronomers took these papers into account. As a Royal Astronomical Society fellow, one hears about spending millions of pounds of taxpayers' money in sending up groups of satellites into space around the Earth, in order to explore the contours of the geomagnetic field, and how they are modulated by the Sun. Modulation by the Sun is OK, they are allowed to discuss that. They are distressed, however, that the public has no interest whatsoever in their work. The public need more 'education', they say. I rather doubt that, and instead suggest that if they were to show a greater awareness of the huge, monthly effect whereby the Moon modulates this field, they would soon find all the public interest they wanted. That modulation is itself affected by the lunar node position. A new science could be here, waiting to be born.

We've been told all our lives that the Earth's geomagnetic field (GMF) comes from iron in the core of the Earth, or maybe to revolving currents of electricity - after all, heat destroys magnetism and it's hot down there. Instead, let's here try to envisage that it is the presence of life on the Earth that is somehow generating this field, maybe by bio-electricity. The current theory gives no hint as to how the Earth's magnetism can reverse its polarity

---

25 Alexander Dubrov, 'Human Biorhythms and the Moon,' New York, 1996, p.69
26 H.J. Lang,' Lunar Periodicity of Color sense of Fish,' *Journal of Interdisciplinary Cycle Reserch*, 8, 1977, pp.317-321 (two of Lang's earlier articles were in German: Dubrov 1996 p.154). For other German studies, see Endres and Schad (ref. 14) p.217.
27 John T. Burns, *Cosmic influences on Humans, Animals and plants*, NJ, 1997, p.55
28 Endres and Schad (ref 18) p.120.

at regular geological intervals, as has assuredly happened, nor does it give any inkling as to how the GMF can vary so much on a day-to-day basis.

The GMF is imprinted with the 27-28 day pattern of the Sun and so varies with that rotation period. The GMF affects all sorts of life-processes on Earth, but one more or less has to go to hard-to-find Russian journals to get information on this subject, because this conflicts with the prevailing paradigm. That prevailing paradigm assumes that the GMF is generated by a merely inorganic process, like the pattern shown by iron filings around a magnet.

Let's try, instead, to suppose that Venus, the Moon and Mars have not got any magnetic fields of their own, because they haven't got life on them. The protective biological membranes around the Earth such as the Van Allen radiation belt make life possible, and they are generated and sustained by Earth's magnetic field. One could say that this is a vitalistic image of what sustains the GMF. You have a kind of choice here, reader, because no-one can prove anything concerning its origin. Both the Sun and the Moon affect the GMF, the former by a pattern which is not rhythmic, not cyclic - an irregular pattern which repeats through the 27-day period of Sol's - whereas Selene's Sphere bestows upon it a monthly wave-pattern modulation. This topic is of relevance to Farmers' Moon because the GMF is often invoked to explain influences, as well as sometimes having a direct effect on its own. I believe that, in the future, the Silver Axioms concerning optimal times for farm activities - maybe functioning as a desktop computer program - will utilize the vector of the GMF as well as the several conditions of the Moon.

Termites orient their nests in an East-West direction, utilizing the GMF, and tree roots likewise orient by it. Birds use it for migration, and during eclipses and geomagnetic storms they can lose their way (the latter, by the way, is more accurately a solar storm, which is impacting upon the GMF).

Fish and especially eels are sensitive to the GMF, and the latter use this to navigate across the Atlantic ocean to spawn in the Sargasso sea. Magnetic activity correlates with the number of fish caught, and all vertebrate marine animals use extremely low-frequency electric fields for orientation when navigating and attacking prey. The human spinal cord is a permanent electromagnet with the negative pole near the head and the positive near the posterior. Deep sleep is improved when the body is oriented in a North-South direction.

Orienting a seed-embryo's root towards magnetic North produces females and towards the South produces males in plants such as hemp and cucumber. Growth tends to be greatest when seeds are oriented towards the South (in the northern hemisphere). For example, pine (conifer) seeds with their embryo roots (radicals) oriented towards the South, germinated four or five days sooner than those oriented towards the North. Decreased GMF intensity (shielded GMF) caused pine seedlings to undergo a prolonged dormant period, as well as reduced seed germination and oxygen uptake.

Likewise this caused barley, pea and millet seedlings' embryonic roots to grow shorter. These comments are intended merely to give a superficial glimpse of a huge but largely ignored subject, where the references tend to be in German or Russian, namely the manner in which living organisms are adapted to the GMF.[29] The GMF somehow represents the mood of Mother Earth. It is a topic which would surely help us to apprehend the nature of lunar influence.

---

29 The references for this paragraph come from R.M. Pasichnyk, 'The Vital Vastness, Vol. 1: Our Living Earth' NE, USA 2002, Ch.15.

10. The Silver Axioms

What time with plough and spade to break the soil,
That plenteous stores may bless the reaper's toil,
What time to plant and prune the vine he shows
And hangs the purple cluster on its bough.

Aratus *Phenomena*

View from Down Under

The pioneering work of Alex Podolinsky amongst Australian organic farmers was well described in 'Secrets of the Soil' by Tompkins and Bird, the follow-up to their bestselling 'The Secret life of Plants' of the 1960s. Podolinsky is of interest as blending the different approaches of Kolisko and Thun, of synodic and sidereal cycles, as used by the Biodynamic farmer. His work has brought new hope, vision and fertile land to the farmers of Australia. Here was his comment about the synodic (Moon-phase) cycle, as given in a lecture to organic growers:

> ... if we expect a drought we may pick to sow towards new moon ... towards new moon you have more digestive activity underground in the soil and that is when you should plough your crops for green manuring, whereas towards full moon you have too much activity in the top of the plant. Our pasture farmers know this only too well. They have a visitor and the farm looks beautiful, the pasture is just glowing, and another visitor comes a fortnight later and it doesn't look nearly as good, apparently for no reason at all. A couple of weeks later it again looks much better. There are periods of change, that is natural. But if, all the time you only have those plants living out of the fertilizer bag you will see no difference. When they begin to look starved you give them more fertilizer and then they look revved up again, but they are not really functioning as plants . There is more water in trees and grass and all plants towards full moon than towards new moon. In the old days good timber cutters chopping down valuable timber would never cut other than towards the new moon. They would not cut towards full moon, the timber was not as good. If we cut hay we also cut as much as possible towards new moon and not towards full moon. We get much better quality hay that way. Towards new moon the plants are much more active underground.[1]

---

1 Alex Podolinsky, 'Bio Dynamic Agriculture Introductory Lectures', Vol. 1, 1985, Fourth Edn. 1990, Victoria, Australia, pp.83,104 and 124.

The polarity he describes between above-ground and below-ground plant activity, as the two ends of a monthly rhythm, is very traditional (it is, we may note, rather comparable to that which Thun has ascribed to her ascending and descending fortnightly periods). His view gives an interesting prediction for when to sow over periods when drought is expected, as happens over the long summer months in Australia. The farmers of Greece and Rome, whose traditions have come down to us in much detail via Pliny, and which were very much based on the course of the lunar month, must have had to face this problem, in a way that British farmers seldom do.

As regards Podolinsky's bold endorsement of the French Napoleonic Code's rule for felling at New Moon, we may note that he has cast his advice into a readily testable form. When researching the field for 'Planting by the Moon', I and Simon Best were not able to locate any experiment that had directly confirmed or refuted this belief. There were plenty of sneering references to it as an example of popular superstition lingering on, and remarks as if experiments to refute it had been performed, but no actual accounts thereof could we locate. More recently however the fine experiments by Ernst Zurcher have indeed validated this ancient notion.[2]

Podolinsky asks his audience to recognize five different kinds of lunar cycle, but this is too many (Chapter 2) and will only tend to bewilder everyone; however he does put a lot less emphasis upon the tropical (ascending-descending) cycle. The recommendations given in PBTM are quite often supported by his advice.

His views as quoted above may not find much by way of confirmation amongst British farmers because the climate here is too variable for such monthly rhythms to be experienced, which is partly why Britain has rather little by way of coherent lunar tradition in farmers' lore.[3] Here is a quote used in PBTM, in a letter from an Australian market gardener:

> We are growers of tomatoes, on a relatively large acreage, and found throughout the years that during the period of the full moon, a noticeable change takes place in the maturing and coloring of tomatoes. This quickening maturity, irrespective of temperature, only takes place two or three days before and after the moon has reached its fullness. During this period, market places on the east and south coastal states have an influx of colored fruit, where in normal times there is a high percentage of green and colored fruit.

---

2 Ernst Zurcher, 'Cosmic Trees and traditional knowledge of lunar rhythms,' 2002. 'Lunar Rhythms in forestry Traditions', Earth, Moon and Planets, 2001 Holland.
3 There were some fine guides in Renaissance England, e.g. G. Markham, 'The whole art of Husbandry', 1631, trans of a continental work originally published in the 1570s; this tradition terminated with the dawn of modern science in the 17th century.

> We have tested this out on many occasions and our statistics over many years have shown more fruit passing through our packing houses during these periods, and that the fruit is much more forward in color.[4]

One would never come across so definite a statement from an English grower! It is an observation which unconsciously echoes the advice given by Pliny the Elder in 74 A.D, as regards when Roman farmers picked their fruit - fruit for storage was picked at New Moon, while that for a more juicy appearance and immediate consumption at the Full. This experience from Down Under directly corroborates the kind of advice that Podolinsky has been giving to farmers. For the cutting and grafting of fruit trees, Podolinsky had an interesting viewpoint, whereby both ends of the lunar month were utilized:

> ... then tie the cuttings up so the air can get to them but not too much, and put them in a cool cellar, or even in the bottom of the fridge. A cool cellar is better, though. In this way we will starve that wood and when it is grafted it will be only too glad to take in new sap.
>
> 'We graft it on towards full moon when a lot of sap is flowing and the piece of wood which has been doubly starved firstly because it was taken towards new moon and secondly because it was starved through putting it into hibernation for a time, will result in a much better 'take'.'[5]

In PBTM we merely echoed Pliny's advice to avoid the Full Moon for such operations. It would be interesting to hear from any gardeners or farmers who have tried out this Podolinsky method. Back in 1976, Thun's Biodynamic gardening calendar, 'Working with the Stars' made an important statement concerning the synodic cycle:

> The influence of the Full Moon, throughout all the years of our research, only brought higher yields when these had been forced by mineral fertilizers or unrotted organic manures rich in nitrogen.[6]

It is rare for Thun to make any reference to the Moon's phase in her work: the gardening calendar based on her researches uses much less evident lunar

---

[4] This letter of 1980 from an Australian grower of 'Pickering Produce' near Sydney, which we quoted in PBTM, was in answer to an inquiry by Dr Geoffrey Dean, who kindly forwarded it.
[5] Podolinsky, ref (1), p.135.
[6] Working with the Stars' 1976, p.6. Her 1986 calendar amplified this theme: 'It is necessary to use well-rotted composts and to look after the humus in the soil. Soil analyses showed repeatedly that from a humus content of 2.5 to 2.8 a soil has the ability to communicate cosmic rhythms to the plant. Too much manure blocks the cosmic influences for the plant. Artificial watering tends to level out differences,' p.18.

cycles such as the nodal cycle, but not the Moon's phase. In the 'Agriculture' lectures given by Rudolf Steiner in 1924, he made the rather magnificent affirmation that: "On days of the Full Moon something immense is taking place on earth", and explained how the transmission of these influences depended on the presence of water.[7]

The central paradox of the Thun calendar, for which it has been much criticized, is that it wholly ignores the one lunar cycle clearly endorsed by Dr Steiner in the lectures which initiated Biodynamic agriculture.[8] Opinion is currently quite polarized on this matter - in the relatively small circles which notice these matters. Podolinsky's view concerning the sidereal rhythm was:

> For large acreage farming and also for any climatic conditions other than where you have enough water or irrigation to counteract the wiles of your local weather pattern, the overriding influence, more important than any reference to zodiac signs, is your local weather pattern. An example of this is wheat which has to be sown into moist soil after rain. If you have had three months drought, as can happen in Australia, and you just have to sow your wheat because the rain has fallen, then that is more important than keeping to a sowing chart. As soon as you have had rain you will sow day and night, never mind what zodiac sign is active. But generally we have found a farmer can still take notice of the sowing chart. Our farmers are so equipped that they can get a huge amount of wheat sown in a day. They just keep going 24 hours a day and they will try to keep to the sowing chart and certainly the wheat is of much better quality if sown correctly according to that chart. The wheat straw is less high when we get our wheat sown under the zodiac sign influencing seed development, whereas if it were sown under a leaf sign it might be twice as high and you have much less head of wheat on the plant.
>
> The market gardeners may have noticed the best beans you have ever picked are on plants that are not huge and have not all that many leaves. They are absolutely full of pods and are deliciously sweet. Even if you let them go a little too long before picking them they do not immediately go stringy. Whereas if you have bean plants that are large and bushy you often have to go looking for the pods because there are so many leaves. People will say, Oh,

---

7 Maria and Matthias Thun, 'Working with the Stars, A Biodynamic Sowing and Planting Calendar', 1978. The Thuns returned to this theme in their 1980 Calendar, concluding: 'To summarize: If plants grow in "living soil," the influence of the full moon is not very noticeable. With incorrect manuring or untimely watering, however, full moon increases yields and decreases quality,' p.20.
8 Rudolf Steiner, 'Lectures on Agriculture,' 1993, pp.23-4.

what beautiful beans you have. But the plant is only leaves and there are not too many pods and those pods shoot too long and too quickly and they are either stringy or they are too watery in flavor. Some of you may have noticed this kind of thing in your own growing of vegetables.

Have you experienced that, and you were never able to understand why that was so? Now if you sow beans under a 'leaf' zodiac sign, then they turn out such huge plants and they have very poor fruit. When you sow under a 'seed' sign, you don't have all that much foliage (the plants don't need it, anyway) but they do have a lot of fruit. For pumpkins, in our experience, sowing under Leo is the most desirable... We have run such trials and we have had roughly four times as many pumpkins in roughly the same acreage sowing them under 'fruit' rather than under 'leaf'.[9]

The Leo constellation, Thun claimed, was the best one for seeds, which seems to be echoed in this Podalinsky lecture, given in 1974. Podalinsky instructs farmers to sow right at the beginning of a lunar constellation (what an astrologer would call its ingress). In his view, there comes a point two days after sowing when a seed breaks down and this he views as a significant moment of chaos: 'When you use our sowing chart it is essential that this germination trial be kept in mind. You must understand that you have to give the seed two full days - the vital unfolding period - within the time that the correct zodiac sign is active.' I never confirmed that and PBTM rather advocated sowing around the middle of the sidereal moon-signs.

Solar eclipses

Karen Hamaker-Zondag, one of Europe's leading astrologers and author of eleven books, put my *Planting by the Moon* calendar to the test. On the same week she and her neighbor sowed radish, she planting in Earth while her neighbor planted in Water (sidereal Moon-sign elements), then they dug up their rows on the same day a few months later. There was a striking difference, her radish being rounded without much leaf, while her neighbor's had gone into long stem and leaf formations. She took photographs showing the difference:

> You see how much difference there is: hers is much longer and it has focussed its growth on leaves, making the radish itself thin. The growth of my radish has focussed on the radish itself, with a much shorter part of the plant above the ground... I have always given this as an example to my students of how your calendar works.

---

9 Podolinsky, ref. (1), p. 123.

Next, she looked at the perhaps equally important nodal cycle:

Another experience: I have planted young pear-trees on the day of a solar eclipse, just to see what was going to happen, because your calendar told me that this is one of the worst days to do so. Well, this was 1984 and they never had any fruit. It took about five years before the first few blossoms appeared, and then last year, for the very first time, there appeared little pears in a beginning stage, but no pear grew big enough to be harvested. Almost all fell to the ground within six weeks after the blossoms disappeared. It is this year, 1993, that I have pears that seem to grow a bit more, almost ten years after planting the trees, and I still have to wait and see if they will become full-grown, but it looks more promising than ever.

This experiment illustrates the enormous value of time-experiments involving the planting of trees, because they are permanent. Any effect from their time of sowing can be observed permanently. In the future, let us hope that we see a university arboretum with trees planted at different times, some over eclipses and others with Moon opposition or trine to the ruling planet, helping students to discern the nature of celestial influence. Karen's letter continued:

Another example: we bought sods because we wanted a lawn. We have tilled the soil and put the sods on it with Moon in water [i.e. a leaf-day]. Well, we'll never do this again, because there is no-one in our neighborhood who has to cut his grass more than we do. It puts all its energy in forming grass-leaves! It recovers fast from damages but it grows very rapidly. Without being specific about their experiences many of my students who worked with your calendar years ago (when I ordered so many of them) were enthusiastic about it, and still some of them would like to have a new calendar.

She was happy with the notion that humans responded to the tropical zodiac, whereas the plant-realm was attuned to the more primordial star-zodiac (she explained to me). In this she was echoing the dualistic view expressed in 1976 by the astrologer John Addey:

It certainly would not be surprising if some of the sidereal points we mentioned earlier... were capable of producing harmonics, and these may refer to terrestrial phenomena as weather cycles, whilst the Tropical reference points provide the basis for the symbolism of the nativities.[10]

---

10 John Addey, 'Harmonics in Astrology', 1986, p.204.

The idea of the malignity of an eclipse is, without a doubt, the most persistent and longest-lasting belief in lunar-gardening (the only competitor in this respect might be the notion that one should not cut down trees during a Full-Moon). Three thousand years ago, the Babylonian tablet Enuma Anu Enlil described the ill-effects resulting from a solar eclipse:

The land of the prince will be destroyed
The land will suffer calamity
The land will go to ruin
Pregnant women will miscarry their unborn.[11]

And Shakespeare put into the mouth of King Lear's advisor quite similar terms:

These late eclipses in the sun and moon portend no good to us: though the wisdom of nature can reason it thus and thus, yet nature finds itself scourged by the sequent effects. - King Lear, Gloucester, Iii.

This alluded to a solar eclipse which passed over Europe in 1605 (with a couple of preceding lunar eclipses) that Shakespeare wove into his story. Is Mother Nature indeed 'scourged' by the 'sequent effects' of a solar eclipse, i.e. that which follows in its wake? The astronomer Kepler gave an explanation as to how eclipses exert their effects, which we quoted in Chapter One. The early Biodynamic sowing-calendars by Franz Rulni advocated not sowing anything important for several days after an eclipse, while its successor the Thun calendar gives just the day of the eclipse as 'no-planting'.

In August 1999 a total solar eclipse passed across southern England. Especially in the month following, September, one read about how British agriculture was in its worst condition since the 1930s, with farmers finding it hardly worth digging potatoes out of the soil, sheep hill farmers going broke and calves and sheep being left abandoned. The causes for this crisis seemed to be quite abstract and hard to grasp: the pound was too strong, or exports to Russia had been lost, etc. This crisis had been building up for some years, with the average age of British farmers at 57 and young persons seeing no future in the profession. Something transformative emerged from this crisis, though, in the shape of a huge increase in the organic food market, the one bright spot on an otherwise bleak horizon.

Some simple experiments demonstrating the effect of an eclipse were done by Theodore Schwenck, and summarized in his classic opus 'Sensitive Chaos': phials of water were shaken in succession before, during and after an eclipse, then seeds were germinated using these phials. They grew less

---

11 The *Enuma Anu Enlil,* quoted in 'From the Omens of Babylon, Astrology and Ancient Mesopotamia' Michael Baigent, 1994, p.103.

from the water shaken during the eclipse. One might have preferred to see the seeds germinated before, during and after the eclipse. But that is how Schwenck did it, assuming that the water had a memory and could retain the impression of the time when it was shaken.[12]

A potato-sowings experiment was done by Mark Moodie, in his garden in Gloucester in 2006, over a solar eclipse. Over each of the seven days 27 March to April 1, 2006, he sowed two rows of potato tubers. These dates were centred around the solar eclipse of March 29th. This was near to the river Severn. A lot of seed potatoes were put into egg boxes, and then for the sowing of each row, potatoes were selected at random. In each row he counted out twenty seed potato tubers. Each day they were sown at 10 am, because that was when the solar eclipse chimed on the 29th. Five months later, he weighed the total yield of the crop per row. Harvesting was around the beginning of August. The experiment didn't quite manage to have each row grown over the same length of time, so the results needed adjusting for this.

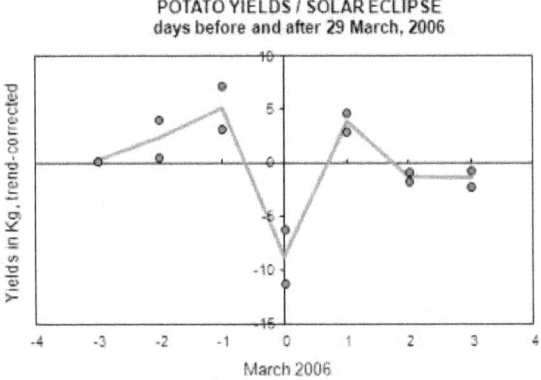

Figure 10.1: The weight in kilograms yield from rows of potatoes sown before, during and after a solar eclipse, trend-corrected for growing period.

Dividing this crop-yield data into two groups gave[13]:

12 rows sown before & after eclipse, mean yield 31.7 ±4 kg;

---

12 Schwenck's graph depicted an inhibition of growth for about an hour centred on the eclipse. Wheat blades in the water shaken at the time of the total eclipse did not grow as high as those of water shaken before or after. His view here was that, while shaking the water, a 'chaos' is introduced which would be receptive to the condition of the cosmos, and which would then remember it as long as it was not re-shaken: 'Sensitive Chaos', 1965, p.66.

[13] For raw data see the *Considera* website: select 'Forums,' then 'Planting by the stars' then 'UK potatoes.' M.M. put that up after the sowing experiment was done, and there averred, 'the difference was clear when the first shoots came up.' There is a link to his photos.

2 rows sown during eclipse: 25.5 + 20.5 kg, mean 23 kg.

The mean value of the two eclipse yield sowings was more than two standard deviations away from the mean of all the other sowings, which is quite impressive. The eclipse-sown rows yielded *almost thirty percent less* in gross yield.

The number of days each row of potatoes was in the ground varied somewhat, between 119 and 132 days, with rows growing for longer having bigger yields. To adjust for this effect, a linear regression line was plotted of days' growth per row, versus crop yield. Then the value given by this line was subtracted from each yield-value, to give a difference value for each sowing date. Thus, we subtracted out the correlation between days of growth and yield from the data to obtain residual values: these were zero-sum i.e. they all add up to zero, they balance out. The graph shows this. As before, dividing this transformed data into two groups gave:

12 rows before and after the solar eclipse: 1.5 ± 2.7 kg

Two rows sown during the eclipse: -11.2 and -6.2, mean -8.7 kg

This time the yield-reduction over the eclipse appears as more than thrice the standard deviation. (About 0.1% of a normal distribution is more than three standard deviations away from its mean.) The data looks better, once corrected for the differing number of days for which each row was grown.

This looks to me like the first decent crop-yield eclipse experiment since historical records began; for example, no experiment of this kind is found in Kolisko's *Agriculture of Tomorrow* (1942). Thun's *Results from the Biodynamic sowing and Planting Calendar* (2003) did allude to sowing trials as having shown a 'decline in the quality of the seed, even going so far as a breakdown in the regenerative powers' over solar eclipses, but no details were given.' One would like to see a two-year trial in a biology lab, to test seed viability consequent upon a solar eclipse.[14]

Cycles of the Sun

Way back in 1975, *Farmers' Weekly* published a report by a cattle farmer Mr Ferrier, who had noticed in his family record book something special about the dates when, in spring, the cattle were put out to graze each year and when they were brought in again for the winter.[15] This followed, he noticed, an 11-year cycle of expansion and contraction, with the cows staying out longest during the sunspot maxima years, as if the grass was growing better over the sunspot maxima.

In the previous year, an article in 'Nature' showed large correlations of yearly root crop yields such as potatoes, turnips and swedes with this 11-

---

14 Mark Moodie and N.K., 'Potato sowings over a lunar eclipse', Star and Furrow, Summer 2012, pp.25-6.
15 J. Farrar 'Sunspot Weather', Farmers Weekly Feb 28, 1975.

year cycle,[16] peaking at the maximum. These articles (by J.W. King) provoked minimal debate or reaction. His employment with the prestigious Rutherford laboratory seemed to end rather soon after and one heard no more on this subject from him.

The subject was an unfashionable one, because, scientists declared, they could not discern by what 'mechanism' such a linkage could exist. The original observations by Sir William Herschel as regards how wheat prices varied with sunspots two centuries earlier[17] had likewise been dismissed. At sunspot maxima every 11 years, wheat prices tended to be lower, Herschel found. Effects on weather and climate became more acceptable and by the late 1970s over a thousand papers had been published on the linkage between sunspots and weather.

As sunspots are dark and relatively cool areas on the Sun's surface, it seemed reasonable to surmise that periods of sunspot maxima were associated with reductions in solar energy emission: was this the link to climate? This question was finally answered in the 1980s, when satellite observations of the Sun discerned that altering solar thermal energy emission was in tune with the 11-year cycle, reaching its maximum during the sunspot peak.

This finding, though paradoxical, at least enabled scientists to understand how the so-called Maunder Minimum had come about: this had been a 'little ice age' period in the seventeenth century, when conditions were so cold that shops were set up on the frozen Thames and oxen were roast upon it. That period saw a spotless Sun, with more or less all sunspots vanishing in the 1640s, a mere few decades after Galileo had first noticed them.

In the 1970s scientists came to accept that the Maunder Minimum period of zero sunspots had been real, and that it could have caused the cold weather. The Maunder Minimum comprised three missing heart-beats of the Sun: a complete solar cycle is 22 years, so that three of these or six 11-year cycles were missing.

In the 1980s, reports started appearing of the 18.6-year lunar-node cycle as present in climate and agriculture records.[18] Experts now regard this as being at least as influential as the solar pulse. The solar cycle may be more evident in more northerly countries, where a greater flux of energized solar particles, as shown for example in the Aurora borealis, can funnel down from the Van Allen belts into the atmosphere. Rainfall or its inverse,

---

16 J.W. King, et al., 'Agriculture and Sunspots', Nature 252, 1974, 2-3.
17 William Herschel, 1801, in Philosophical Transactions of the Royal Society, London, pp. 265, 354.
18 Louis M.Thompson, 'The 18.6-year lunar cycle: Its possible relation to agriculture', Cycles March/April 1989, pp.64-69; 'the 18.6 year cycle in the general economy' Cycles May/June 1989, pp.139-141.

drought, is linked with this so-called 'nutation-cycle' in several parts of the world, e.g. the monsoons of India, as are tidal waves in the Earth's oceans and atmosphere.

In times to come, let us hope that these two long-term cycles will come to be used in economic forecasting, the 18.6-year node cycle and the 22-year sunspot cycle. In both cases one-half of the cycle can appear as being the key length: 9.3 years for the nodes and 11.1 years for sunspots. 9.3 years can be viewed as the eclipse cycle, insofar as eclipses take that long to revolve around the seasons of the year, there being no difference between the north and south nodes in this regard; whereas the cycle of lunar 'declination,' how high it rises in the sky, varies with the 18.6 year nutation cycle.

In the US, high yield periods nearly free of drought tend to occur near minimum declination. These important findings by Robert Currie were generally published in the now-defunct US journal 'Cycles'.[19] Major US droughts, he argued, remain in step with the 22-year cycle rather than the 11-year cycle.

A 9.3 year cycle in US grain production and prices was discussed by Thompson, with high corn yields occurring in years of minimum declination. Also, lowest agricultural prices tended to occur every 18.6 years. Currie has also argued that a crop production cycle of 18.6 years exists. Earlier on, Dewey had identified approximately 9 and 18-19 year cycles in wheat prices.

Periods of 'global warming' may be more related to solar cycles, than global carbon dioxide emissions. I posted the following onto my lunar-gardening website, under 'Forget Global Warming':

> Mother Earth holds onto the balance, her equilibria, so that the weather remains neither too hot or too cold, too dry or too moist: our lives continue and the oceans have never boiled dry or frozen over. Over this last year, Earth's mean surface temperature has dropped by more than half a degree [from Hadley Centre for Climate Prediction], which may not sound much but it means – forget about Global Warming! The Antarctic is now stacked up with record levels of sea ice, while North America this winter (2007/8) is enjoying the most snow cover in fifty years. So, don't worry about the penguins. Last year's precipitous drop in

---

19 R. Currie, 'Evidence for 18.6-year Signal in Temperature and Drought Conditions in North America since AD 1800,' Journal of Geophysical Reserch, 1981, 86, p.11055; 'Evidence for 18.6-year lunar nodal drought in Western North America during the past millennium,' J.Geophysical Res., 1984, 89, pp.1295-1308; 'Examples of 18.6-year and cyclic 11-year terms in World Weather Records' in: 'Climate: History, Periodicity and Predictability,' Ed. Rampino et al., NY 1987.

temperature has clean wiped out that 'global warming' trend, and is making weather scientists realize instead just how dependent we are on the grand cycles of the Sun. Reduced solar activity seems to have led to this drop. That is a much larger driver of climate change than man-made greenhouse gases.

Before global warming, everyone was fretting about global cooling and the coming ice age, remember? Rather, we need to experience Gaia as a living being whose pulsating rhythms are deeply attuned to the Sun and Moon. Instead of all this anxiety, the world needs futurology institutes that will study the majestic and glorious rhythms of Sol, which are the heartbeat of our solar system, and the rather complicated ways in which these affect human activity, climate and agricultural yields on Earth; as well as (this is the difficult bit) separating out the quite large effects that result from the 18.6 year moon-node rhythm."[20]

A survey of over two hundred years of German wines from around the Rhine showed that their best years tended to concentrate over sunspot maxima.[21] From 1848 to 1915, years of excellence in Beaujolais wines have shown the same tendency, according to a report cited by Kolisko.[22]

The Hour of Moonrise

In the 1930s, when Kolisko was reporting her results, various other people also reported that waxing Moon sowings were preferable to those in the waning Moon. Our recommendation returns to this adage, hoary with age, but from a rather different point of view: the hour of Moonrise only occurs during the hours of daylight, only during one-half of each lunar month - the waxing half.

It was shown by biologist Professor Frank Brown that the metabolism of root vegetables kept in the dark at uniform temperature rose to a peak at moonrise.[23] A comparable effect may exist for crops sown then, in relation to their final crop yield, i.e. Moonrise may be the optimal time of day for sowing a crop. This effect was observed in trials conducted in the open, in contrast with the previous report which dealt with in vitro seed germination.

---

20 www.plantingbythemoon.co.uk
21 A. Rima, 'Consideration su una serie agraria biseclare: la produzione di vino no Rheingau 1719-1950', Geofis. e Meteor., XII 1963, p.29; cited in Gauquelin, 'The Cosmic Clocks' p.100. Gauquelin also stated that the French Astronomical Bulletin had shown the same effect for Burgundy wines, but gave no reference.
22 L. Kolisko, 'Agriculture for Tomorrow', p.3, citing work in Lakhovsky.
23 Frank Brown, 'Biological Clocks: Endogenous Cycles synchronized by subtle geophysical rhythms,' Biosystems (Amsterdam) 1976, 8, p.75.

Colin Bishop's 1978 radish trial was described in chapter 6, when for 39 successive days he sowed four rows every day, two in the morning on his way to work and two more in the evening on his return.

He noted the time for each sowing, and also watered each row prior to sowing the seed, so that they began germinating upon being sown. The main monthly cycle there present appeared as the 27.3-day sidereal cycle, not the synodic period.

This supported the findings of several other investigators, that what is effectively a third harmonic of the sidereal lunar month influences crop yield, timed by the Moon's passage against the elements of the sidereal zodiac. Sixteen years after the experiment was performed, I finally discovered how to perform a proper analysis of the data. It required a multivariate analysis, subtracting out the different components one at a time.

Lunar Day

In 1978, Best presented a review of all evidence regarding lunar influence upon plant growth. He concluded that the rhythms displayed the first four harmonics of the lunar month, with the third harmonic as primarily working through the sidereal month, in phase with the Chaldean zodiac divisions. These findings endorse such a perspective.

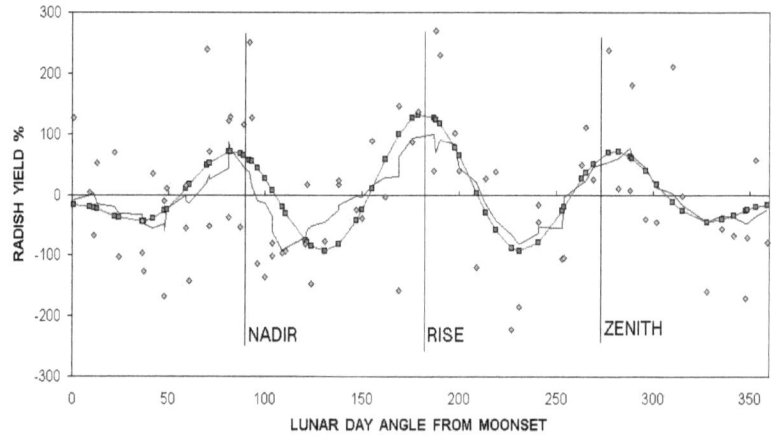

THE LUNAR DAY
SHOWING A WAVEFORM AND MOVING AVERAGE

Fig 10.2 depicts the data plotted against the 24.8-hour lunar day. This set of data is unique in having a large number of timed sowings and no rows omitted or spoilt. It gives an opportunity for resolving issues that have long remained embedded in folklore. For the best-fit sinewave equation here plotted, see Appendix 4.

A Victorian astrological journal[24] used to run a regular gardening column, which would confidently recommend the two hours centered on moonrise for sowing seeds etc. The results presented here support this tradition.

Moonrise is latitude-dependent. Therefore, the results here reported will tend to encourage the use of a computer program able to define optimal times for the sowing of crops: a kind of farmer's watch, set to lunar time. Organic growers will thereby be able to tune into the living rhythms of the Earth for optimizing the growth of crops.

GMF

Earth's geomagnetic field (GMF) responds to solar radiation. On average the sunspots revolve around the Sun's face once per 27.3 days (a figure you have heard before) and accordingly this period is found in the geomagnetism indices. So, if a 27.3 day period is found in, say, crop yields, there are different places it could be coming from! It could be solar or lunar or terrestrial, if perchance geomagnetism was viewed as affecting crop yield ... the two luminaries in the sky, of the same angular size, may both be said to revolve once per 27.3 days as observed from Earth, a curious situation. The solar magnetic field streams out from the Sun and revolves with it, being normally divided into four sectors. As the sector boundaries sweep past the Earth, in this 27-day period, climatic phenomena such as thunderstorms and lightning are linked thereto.

As to whether a higher GMF at sowing time is good for plants, the diagram shows a regression line put through the Bishop data. It indicates a modulation in crop yield, as a function of the daily-changing GMF. Its magnitude was ± 28% (the mean value here is 239, oz. x 100). Another diagram shows the Spiess radish-yield data plotted by local GMF at sowing date. This is a fairly modest effect of around 10%.

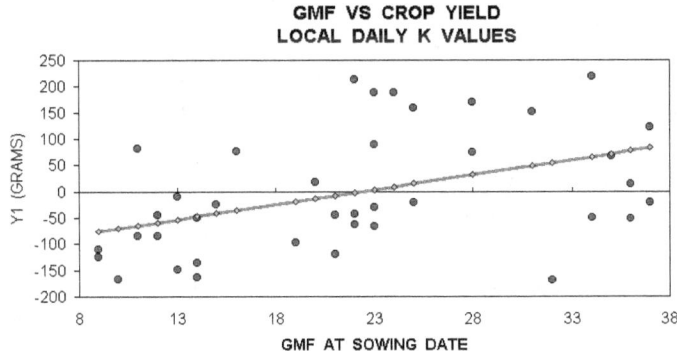

24 'The Astrologer,' a Victorian astrology journal of London.

Fig 10.3 The Bishop radish data (1978) with the geo-magnetic field at sowing date plotted against yield.

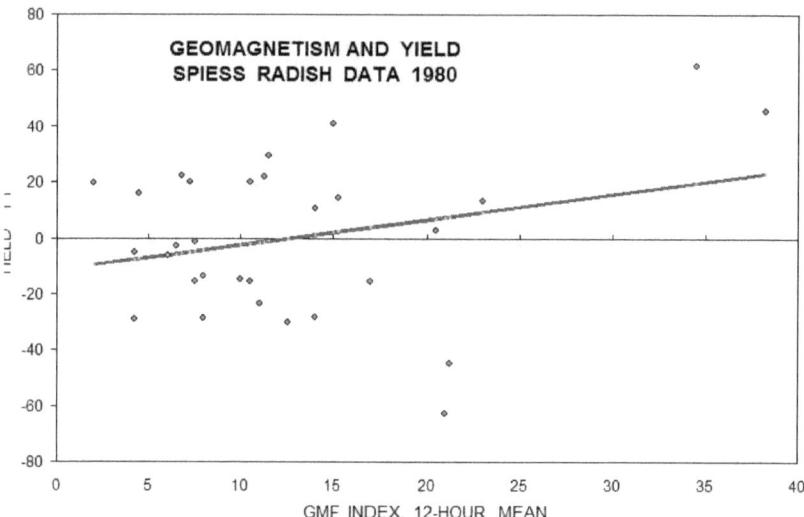

Fig 10.4 I plotted the yield of Spiess's 1980 radish data against GMF field strength at time of sowing.

A couple of books may be recommended concerning geomagnetism: one is that old classic *The Geomagnetic Field and Life* by Alexandre Dubrov, translated from the Russian, and the other is *The Vital Vastness* by the young writer Richard Pasichnyk (2002). As Dubrov well explains, there are two parameters of the GMF liable to vary: one is its strength and the other its direction as a vector. Both of these fluctuate remarkably, indeed if you believe that a large iron core is responsible for the GMF then it may be quite an eye-opener to see just how much these vary from day to day.

By way of encouraging a more vitalistic attitude towards this enveloping field that surrounds Mother Earth, here is a fairly extensive quotation from the latter book (I've here taken the liberty of removing all references from the text):

> Plant growth and respiration are intimately related to the Earth's magnetic field. Plant respiration, growth, biochemical reactions, algae reducing nitrate, metabolic variations and other physiological reactions are synchronized with cycles in the geomagnetic field (GMF). Decreased GMF intensity caused pine seedlings to undergo a prolonged dormant period, as well as reduced seed germination, oxygen uptake and dry matter content by 30%. Under similar conditions mustard plant growth was retarded. Likewise, barley, vetch, pea and millet seedlings'

embryonic roots were shorter, while the plants also underwent various changes in growth. When exposed to a magnetic field stronger than that of the Earth's, dandelions flowers delayed their opening and closing, and when a more lengthy exposure was enacted they wilted and died.

A plant and its root 'align with the Earth's geomagnetic field. The side roots of beets, for example, were noted to grow along a uniform compass direction. When a plant embryo is oriented toward the North or South pole, different growth responses occur. Depending on the plant, growth was greater in one direction than the other. Red silk-cotton flowers displayed different shapes (dissimetry) according to the geographic location and its relationship to the angle (dip) of the GMF. A similar phenomena was observed with sugar beets, corn, wheat and radish root-creases (sic), and overall plant response and separation into new functional types. Orienting the seed-embryo's root towards magnetic north produces females and toward south produces males in plants, such as hemp and cucumber. Space, germination, growth and root development of any plant is oriented with respect to the GMF. Growth is greatest when seeds were oriented towards the South Pole in the Northern Hemisphere...

Growth response according to the GMF is referred to as geomagnetotropism and can be found in all plants in one form or another. For example pine (conifer) seeds with their embryo roots (radicles) oriented towards the South, germinated four or five days sooner than those oriented towards the North.

Lunar phase is known to affect the Earth's geomagnetic field and electrical environment, and likewise, plants respond to lunar phase. The levels and rhythm of root excretions by plants depends on the time of soaking in relation to lunar phase. Growth was at its best during Full and Quarter Moon phases. Undoubtedly a strong and stable geomagnetic field is the best environment for the healthy growth of plants, and plant abundance strengthens the GMF by producing magnetic, electric, electrostatic and electromagnetic fields...

Insects respond in various ways to both electric and magnetic fields. They orient in flight direction to both electric and magnetic fields...Bees can more accurately communicate the location of pollen bearing flowers using the GMF as a reference frame. A sharp increase in the weight of beetle larvae takes place during unusually strong disturbances in the geomagnetic field. Even attraction to a lamp or light, depends on the state of the GMF.

During magnetic storms the daily (diurnal) rhythm is greatly altered, when great numbers of insects take to flight...

All insects react, though somewhat differently, to the GMF and magnetic disturbances. For example, termites orient their nests in an east-West direction. Rhythms in the termite's feeding behavior correlate to magnetic and solar activity. This is also true of the feeding behavior of some beetle species. The Earth's electrical environment influences the activity of flies, and insects' growth, development from larvae, shedding and egg-laying.

GMF orientation is involved in fish migration, and as a result, data show a correlation between magnetic activity and the number of fish caught. ... The earth's magnetic field is utilized so well by birds that they can be blindfolded and still reach their destination. In contrast, during eclipses and other times of disturbance in the earth's magnetic field, such as geomagnetic storms, birds lose some or all of their homing capabilities...

For humans, 'something as simple as orienting the body at a 45° angle in relation to the main compass points can reduce the amount of deep sleep. An improved sleep and psychobiological well-being are achieved when the body is oriented in a north-south direction.[25]

Pasichnyk's opus has a commendably vitalistic and holistic approach, but is also optimistic, whereby Mother Nature is seen as self-organizing in a manner that maximizes biodiversity: as part of which the GMF appears as very much formed by living processes on Earth, which could be why Venus, Mars and the Moon around the Earth have no planetary magnetic fields to speak of. And, why dandelions do not grow so well if nurtured in a magnetic field stronger than Earth's.[26]

Equations

Mathematics is a universal language. We here write the time-equations of celestial influence, never seen before. The Silver Axioms are translated into mathematical form. It is not necessary for farmers and gardeners to follow these equations: they will go into the 'Farmers' Moon' computer programs of futurity. The busy farmer will just experience them as optimal time-recommendations.

We adhere to the sequence of the earlier chapters, expressing the results as continuous solar-lunar wave-functions. We write a sine-wave as A sine (S-

---

25 Richard Pasichnyk, 'The Vital Vastness Volume 1, Our Living Earth' 2002, pp.163-178.
26 A study showed that potatoes grew better when grown from 'excised magnetically treated eyes' of seed potatoes then others grown from eyes not so treated: U. Pittman, 'Biomagnetic response in potatoes', Canadian Journal of Plant Science, 1972, 52, pp.727-733.

M), S-M being the difference between solar and lunar longitudes, and A the amplitude, i.e. how big the effect is. This sine-wave will peak at 90° of longitude, i.e. the first lunar quarter, then will minimize half a cycle later. A Full Moon position happens when 180° of celestial longitude separates the two luminaries so they stand opposite each other in the sky. To make the sine-wave peak at this Full position, we adjust the function by 90°, or else use a cosine function: sin (S-M-90°) or -cos(S-M). Maybe sketch out the old sine and cosine waves, like you did long ago at school. Farmers and gardeners *do not need* to understand these matters, they only need the instructions on how to use the Farmer's Moon clocks that will (let us hope) sit on the desks of sensible organic growers in futurity – based on these equations. We are here dealing with *time-equations*, which have a *universal validity*. They describe how rhythms of life are connected to the luminaries in the sky, whereby millennia-old calendar-rules are re-envisaged as continuous functions that ebb and flow.

We start with the Brown and Chow bean seed water absorption experiment of Chapter 3, where seed water absorption peaked symmetrically at the four lunar quarters:

$$W/W_m = 1 + 0.06\cos 4(S-M) \qquad \text{(Fig 4.1)}$$

where W is water absorption and W_m is the mean value (i.e. absorption over two hours at a given temperature). The amplitude variation of the water absorption was about 6% (you may wish to turn back to Chapter Three to see the graph), while the factor of 4 indicates that it will go through four cycles per lunar month. It will peak at the Full and New Moon positions as well as the quarters.

Chapter 3 also looked at tree cell mitosis, (as reported rather briefly by Vlasinova et al. in the journal *Biologia Plantarum* in 2003). About 8% of the embryo tree cells investigated (in vitro) were dividing at any given time. This fraction they found had a distinctive lunar rhythm, which was a sum of two waveforms, the first and second harmonics of the lunar month. Thus there is a solar-lunar rhythm of cell division in trees.

The investigators kept tree seedlings in a laboratory, and they gave the amplitude of the two waveforms they found, which enables us to write the equation: it's a tidal-type rhythm, i.e. fortnightly, peaking at the Full Moon. The diagram shows how the two separate waveforms add up together.

We are startled by how large the effect is, varying from 8.8 % cell division (i.e. 'mitosis' – that is the percentage of tree cells that are dividing at any given time) at the Full down to 7.5% at the two quarters. This means that the rate at which cells were dividing in a laboratory varied in the course of a lunar month (over the three months of their experiment January-April 2001) by 16%. Just mull over that figure for a while, and what it means. It's part of a wonderful song, of the trees.

The study by Oehmke of bee activity showed a tidal, bimonthly fluctuation in numbers, as they moved in and out of their hives, but its phase had a curious seasonal shift:

> In Morocco, the native honey bees showed a semi-lunar flight activity during the summer months with maximum values during the quadrature. From the end of August to the beginning of september a transition phase was observed in which the activity peaked during the syzygium.

We write these two fertility-equations as:

Summertime: $B/B_m = 1 + 0.2\sin2(S-M)$

Autumn: $B/B_m = 1 + 0.2\cos2(S-M)$

where B represents the frequency of bees approaching and leaving the hive. Oehmke's graphs indicated that the amplitude of this effect was around 20%. The first of these equations peaks at the quadratures, the second at the syzygies. Nothing is more important for the fertility of Mother Nature than bee activity.

We are here seeing a subtle fourfold architecture vis a vis the lunar month in this most geometrical of insect species, possibly reminding us of its hexagonal hive lattice, or its figure of eight dance whereby it explains where the honey is.

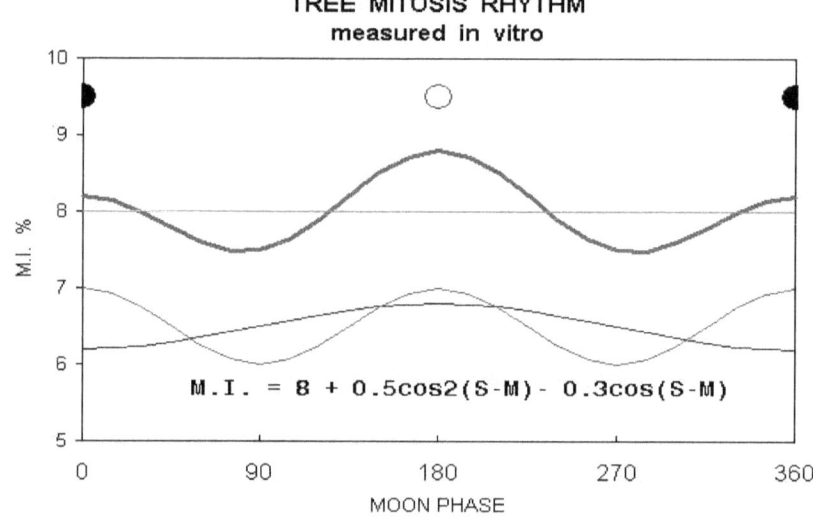

Fig 10.5 Tree-cell mitosis rhythm (in vitro) plotted against lunar phase: Vlasinova et al., 2003.

This writer's wheat-seed lunar-germination trials obtained only a 7% amplitude, peaking at $S-M = 130°$, that is about four days before the Full

Moon (Luna moves 13° along the zodiac every day). We may write this equation as:

$$L/L_m = 1 + 0.07\sin(S-M + 40°) \qquad \text{(Fig 3.5)}$$

where L was the total length of wheat grown over one week, in the dark at uniform temperature, Lm being the mean value. That trial was done indoors at constant temperature. Kolisko in her seed-germination trials of the 1930s consistently found that seeds germinated optimally two days before the Full Moon, and least well two days before the New Moon. My experiment done indoors and in a growth-chamber shows a smaller-amplitude effect – but, with a peak in the same position.

The sidereal rhythms here examined were a function of lunar longitude. We can tentatively represent the 'classic' Thun result, her three years of potato trials 1963-5 (Chapter 4), as:

$$Y/Y_m = 1 + 0.15\sin 3(Ms - 15°) \qquad \text{(Figs 4.3-4.5)}$$

where Y is the crop yield, Ym is the mean crop yield, and Ms is lunar longitude, measured in the Sidereal Zodiac. The number 3 makes this expression go through three cycles per sidereal month. For the waveform to peak in Earth we subtract 15°. The element-sequence starts from zero degrees of Aries: Fire, Earth, Air, Water.

The horse fertility data of Chapter 9 showed an interaction between two different waveforms: the endogenous 21-day estrus cycle and the exogenous lunar-monthly period, the latter being modulated by the 18.6 - year lunar-node cycle. The first of these we may write as:

$$F/F_m = 1 + 0.25\sin(360° \times T/20.5) \qquad \text{(Fig 9.1)}$$

where F is horse fertility (likelihood of mare conception per covering), F_m is the overall mean mare fertility for the breeding-season, T is time in days counting from when a cycle begins. That will cycle once per 20.5 days, the estrus cycle.

The amplitude of this effect (which was about 25% in the 1990 data, Ch. 10) may depend on how close together the mares are, sharing pheromones that get them 'on heat', e.g. whether they share stables at night.

We saw how the pattern of horse fertility in the lunar month reached a minimum at or near to the first quarter and a maximum near to Full, unsuspected by millennia of folk-tradition.

A mere nine or ten days separated the peak and trough within this monthly cycle. A single sine-wave cannot model this, we require two at least. They are summed, that is added together, just as we did for the moonrise equation.

To create the figure 10.6, depicting horse conception in the lunar month, we start off with the total number of registered conceptions – as I copied them down at Newmarket - for the complete data-set. We plot them by their lunar-day number on the day of mare-stallion copulation, counting from 1 at the New Moon and so reaching Full at or around 14.

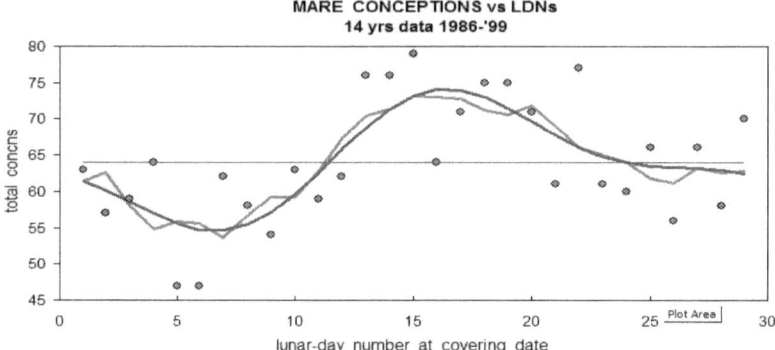

Fig 10.6 The horse conception data summed by lunar-day numbers, plus a moving average (five-day) and a best-fit sine function.

The diagram also shows a curve giving a good approximation to the data, made by combining two sine-waves, the first and second harmonics of the lunar month, as:

$$C/C_m = 1 - 0.12\cos(360° \times (T-4)/29) + 0.05\cos2(360° \times (T-8)/29)$$

where C is the number of registered conceptions and T is the lunar-day number $1 - 29$.

The graph thus compares this waveform, the sum of two sine-waves, with a five-day moving average. That waveform has *a thirty percent differential* between its peak on day 16 and its trough on day 7. That is a universal statement: that, in all corners of the globe, a thirty percent differential or thereabouts will exist between the number of mare conceptions around the Full Moon and that nine days earlier in the first quarter of the lunar month. Two waveforms of 12% and 5% amplitude are summed to give this graph, the second of which is a second harmonic, i.e. it goes through two cycles per lunar month.

That is an average effect as it appeared over 14 years. Then the amplitude of this effect is modulated according to the lunar-node cycle of 18.6 years. For comparison, we saw how geomagnetism was strongest at the Full Moons which lay nearest to the ecliptic i.e. near to the lunar nodes. In contrast horse fertility was (surprisingly) maximal at Full Moons furthest from the ecliptic, i.e on those Full Moons having maximal celestial latitude.

These equations give expression to the Silver Axioms, which link together Heaven and Earth. Appendix 4 gives some more details.

Lynx and the lunar node

By way of indicating future prospects, let's end by looking at lynx-pelt abundance data. Sales of the Canadian lynx pelt were recorded by the Hudson Bay Company, and John Dewey, founder of the Foundation for the Study of Cycles in Pittsburgh,[27] found yearly data over the period 1735-1960 - that is a long, continuous record over two centuries.

His diagram, published in 1973, indicated a huge, uninterrupted rhythmic oscillation in lynx pelt abundance over this entire period, and Dewey estimated its period as 9.6-years. This fluctuated from under 2,000 skins on a poor year to over 70,000 in a good one. Dewey further averred that this 9.6 year period was:

> ...characteristic of much wildlife: The coyote, red fox, fisher, marten, wolf, mink and skunk have abundance cycles of the same period.[28]

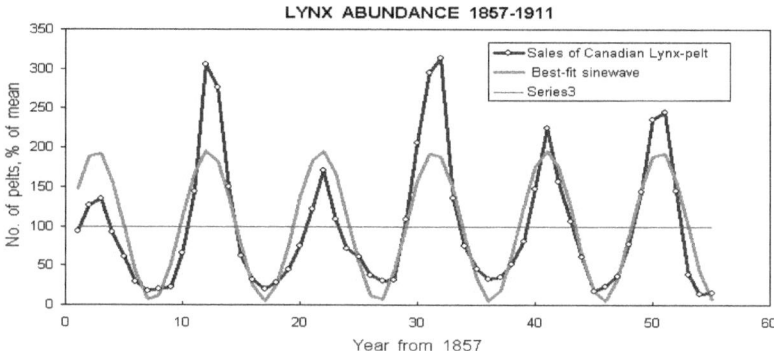

Fig 10.7 Lynx abundance, as shown by sales of pelts in Canada, showing massive 9.6-year periodicity with clockwork regularity over fifty years, cause unknown (Data from Dewey, 1973).

and he also discerned this same period in the abundance of Atlantic salmon. It was present both in British and American data but more strongly present in the latter. It is indeed of interest if this rhythm was more strongly present in the US than in European data. Could this period, we wonder, be related to the 9.3-year half-node period?

---

27 The Foundation for the Study of Cycles at Pittsburgh continued after Dewey's death, but became more narrowly focused on market cycles and went defunct in about 1996. In 2004 the Cycles Research Institute was founded, carrying on this work.
28 John Dewey & Og Mandino, *Cycles, the Mysterious Forces That Trigger Events*, 1973, p.26.

I found a more limited data-source, for only 54 years of the Canadian lynx-pelt data (1857-1911), and this clearly showed the same rhythm as had been discovered by Dewey: it had a staggering 95% amplitude! We write its equation[29] as:

$$Ly/Ly_m = 1 + 0.95\sin(360°\times(Y-p)/9.6) \quad \text{(Fig 10.7)}$$

where Y is the time in years and p an arbitrary starting-date. I verified from this data-set (see figure) that its period was indeed 9.6 years as Dewey had said, not 9.5 or 9.7. How can a biological rhythm continue with such a huge magnitude? Where is the clock? Out in the mountains, the Canadian lynx (or lynx pelts) were increasing and decreasing by a factor of ten with metronome regularity every 9.6 years! One only wishes that biologists were interested in such a topic.

---

[29] Here the equation involves sine/cosine functions of a time-variable in days or years (lynx-abundance, horse conception, tree-bud rhythms), whereas others involve angular measure, viz the Sun-Moon angle.

Conclusion

That which is above is like to that which is below, and that which is below is like to that which is above, by which means the miracle of unity is accomplished ... With great sagacity it ascends gently from Earth to Heaven. Again it descends to Earth, and unites in itself the force from things superior and things inferior. Thus will you possess the glory and brightness of the whole world, and all obscurity will fly far from you. - Tablet of Hermes, Syria, 4th century.

There is a cosmo-agriculture of tomorrow which is ready to be born, able to bring dignity into the life of the farmer, by placing him or her to some degree between heaven and earth – in accord with the vision of the Hermetic tablet. The perspectives here involved can bring a deeper appreciation of the matrix of life, that within which life evolved, as it is woven between the interaction of Earth, Sun and Moon. More work on this whole area needs to be done, not least because of its potential benefits to countries in the third world whereby they might increase the quality and yields of their crops, cut back on imports of environmentally damaging agrochemicals, *plus* also help to maintain their traditional beliefs about the world. In tropical countries where drought can be a severe problem, planting around the Full or New Moon may turn out to be of paramount importance, as biologists have shown water uptake to be linked to these times of the month, as too is rainfall. As agricultural research stations around the world tire of new wonder-chemical and genetic treatments, which turn out to be rather more harmful than had been supposed, perhaps they will look anew at the mysterious attunement of plants to the cycles of the Moon.

Plants, we have here argued, do not have an 'inner life' in terms of a heartbeat and their own bodily cycles, they are more open to the cosmic processes, and are therefore much more responsive to it than we are. We may also have apprehended how all living things respond to the synodic lunar cycle, especially in the matter of fertility. In this 21st century a new science of chronobiology will comprehend these things, which may wish to refrain from cutting open small furry creatures. The Silver Axioms, as here described, may come to be used in the lunar-fertility computer programs of futurity. They represent an intertwining of synodic and sidereal rhythms, and have a practical importance for the farmer. The role of geomagnetism will also come to be recognized as important here, though its role is not yet well understood, and such programs will need real-time input of the local GMF. The present work has made predictions that are testable, even though such tests may not be quite so cut-and-dried (let's reflect that this phrase means, no longer alive) as that of traditional science.

We saw how, from horse-breeding data in the stables of Newmarket, there were substantially more live births from all coverings over the second lunar quarter than over the fourth. It looked as if the node and synodic cycles were here interacting. We have not been in a hurry to find an 'explanation' for this phenomenon. There is no mechanism, no process 'behind' it, to account for it. Eventually these equations will be presented at horse fertility seminars, taking their minds away for a moment from stallion testicles and chromosome structure, where one should just let Mother Nature do her stuff.

Mother Earth seeks out her optimal equilibria. The astronomer Johannes Kepler formulated what we here are meaning by Gaia, the living Earth. His *De Fundamentis* of 1602 described how the Earth-soul responded to the condition of the heavens in a geometrical manner, because Earth was a being that was sentient, even though it was not conscious.[1] We are heirs to his perspective, ours is a vitalistic view. The word 'Gaia' was coined by the British atmospheric scientist James Lovelock, who envisaged how Earth's atmosphere adjusted as if the Earth were somehow alive. However Lovelock's view of what he meant by 'alive' turned out to be entirely Darwinian, and his Gaia was a blind and purposeless entity. Her balances were nothing but feedback mechanisms![2] We seek for a very different experience of Mother Earth, and let's here quote the beautiful words of forestry professor Ernst Zurcher, trying to express the significance of his findings:

> On this basis, a fruitful exchange is possible between scientists and foresters who are aware of the 'cosmic' dimension of trees and its philosophical/scientific meaning. As a matter of fact, the works presented here on astronomic rhythms in organic life give an insight into an unexpectedly common level between trees and human beings. They lead to a rehabilitation of parts of ancient, almost forgotten knowledge. One positive consequence is the enhancement of the intrinsic value of each tree, from a physical, and also a social and spiritual view.[3]

All living things rejoice, so to speak, in rhythmic activity and we have here focussed upon those that are harmonics of lunar periods, and which have a practical meaning for agriculture. But, in conclusion, let's glance briefly at one having a purely theoretical interest: a fish swims in a bowl, and its color

---

[1] Johannes Kepler, 'On Giving Astrology Sounder Foundations, A new short dissertation concerning cosmology, with a physical prognosis for the coming year 1602,' translated in Judith Field's 'A Lutheran astrologer: Johannes Kepler', *Archive for History of Exact Sciences,* 1984, 31, pp.190-268

[2] James Lovelock, *The Ages of Gaia, A Biography of Our living Earth*, 1988.

[3] Ernst Zurcher, 'Cosmic trees and traditional knowledge of lunar rhythms' *Moving Worldviews* Ed. Haverkort and Reijntjes, Leusden 2006 .

sense varies with the lunar month. Various fish species are found to have a more intense yellow perception around the time of the Full Moon, and then to have a better perception of violet two weeks later around the time of the New. This shift in color perception was found to be common for all species examined, and it is not impossible that this experience is common to fish in the sea.[4] Human color sense also varies with the lunar month, with a better red perception at the Full Moon and less blue[5] and it seems a pity that modern biologists are not interested in this topic. It is a natural phenomenon, i.e. it's what Nature does.

Could we not have a more peaceful culture, with less violence, and more in touch with feminine values, if we learnt to accommodate ourselves more, and tune into, these solar-lunar life-rhythms?

We speak of leading someone 'up the garden path' as meaning to mislead them, as if there were leprechauns at the bottom of the garden ready to play some prank. But, can one hope to comprehend the mysterious unity of things, the delicate web of life, in a laboratory with its fluorescent lighting and machinery? One may need a new phrase for the damaging effects of the reality-concepts inculcated by textbooks: "Why, you are leading me into the laboratory!" as meaning, long-term environmental damage, harmful side-effects, spiraling costs, illusory mechanistic reality-concepts, rising cancer mortality levels and loss of all human meaning. We have here valued those experiments that are simple, and not too far removed from what Mother Nature herself does.

One would like to see a laboratory where the apparatus was simple enough to give delight to a child. In one corner of this lab would be a series of phials such as Maria Thun prepared, containing sunflower oil. They would have the total oil extracted from the same quantities of seed sown in the four different sidereal elements, so that students could inspect the different smells, colors and quantities of oil. Would the oil from the air (flower) day sowing be the best, or perchance would it be the warmth/fire day (seed)? Then, would the optimal times for sowing sunflower fields be in the hours before a Moon-Jupiter trine or opposition? Outside the laboratory would be two rows of apple trees, one planted (or grafted) during a solar eclipse and the other at an 'optimal' time. An apple tree should be planted as Venus is rising, in consideration for the tree. Tests would be performed to inspect any differences.

Slicing open a red cabbage, one discerns its formative pattern, and whether it looks healthy or not.

---

4 H Lang, 'Lunar Periodicity of Color Sense of Fish', Jnl. Interdisciplinary Cycles Research 1977, 8, pp.317-321.
5 For color perception in the lunar month, see Dubrov, Human Biorhythms and the Moon NY 1996; Endes and Schad, 2001; and Klein, 2007, p.26.

Maybe our laboratory would have a scale model of a cow, in perspex, containing gold coloration to model the varying concentration of gold within it. This would show how the highest concentration of gold is reached in its horns. This might help Biodynamic farmers to appreciate why it is so important for their cows to keep their horns on, which is not achieved in normal British organic farms. It might also remind us of the golden solar disc which Egyptian artists once inserted between a cow's horns. The horns are the one part of a cow that reaches upwards, and they give dignity to the cow.

On the other side of the lab would be a life-size human in Perspex, showing a similar gold-gradient, indicating how the peak gold levels are here found in the region of the heart. Gold is not as such biochemically active, so these levels are a bit of a mystery.

Rudolf Steiner envisaged the first movement for organic agriculture and et's have a few quotes from him about what he meant. He said:

> A farm is true to its essential nature, in the best sense of the word, if it is conceived as a kind of individual entity in itself – a self-contained individuality. Every farm should approximate to this condition. This ideal cannot be absolutely attained, but it should be observed as far as possible. Whatever you need for agricultural production, you should try to possess it within the farm itself.[6]

This is what would nowadays be called a holistic philosophy. He characterized the farmer's life as a meditative one. Today, pressure is being exerted upon those who control the definitions of organic agriculture, as supermarket chains, complying with the public's wish, seek to open up huge 'organically grown' vegetable markets.

The premier holders of these definitions are Demeter (Biodynamic) and the Soil Association, and let's hope they are strong enough to hang onto what they believe organics should be. The concept of organics needs to be linked with slow development and small-scale farming. 'Demeter' has long required seven years of non-use of pesticides etc before they will give their much-desired seal of approval.

While advising farmers about to set up the original organic-farming movement, Steiner referred to earthworms as:

> Wonderful regulators, safety valves for the vitality inside the earth. These golden creatures – for they are of the greatest vale to the earth – are none other than the earthworms."[7]

That is a meditative statement, which can be mulled over, pertaining to the thin layer of topsoil 'humus'. Its condition seems to be critical for the

---

6 Rudolf Steiner, '*Lectures on Agriculture.*'
7 Ibid.

Conclusion

response to lunar/celestial influences. Earthworms are continually being drummed out of the topsoil every time the agribusiness agent (let's not call him a farmer) sprays poison over the land.

In Chapter 1 we saw the staggering fact that a survey in New Zealand found that the soil of Biodynamic farms had twenty times more earthworms, than did the soil of neighboring farms. We have seen how some general indications which Dr Steiner gave for cosmo-agriculture were a stimulus for the Biodynamic calendars to develop.

Here is a prophetic comment from Steiner concerning automation and the life-processes of Mother Nature:

> When modern technology has made it possible to warm large areas with artificial heat – I am not finding fault but merely telling you of something that will necessarily come about in the future – then plant growth, above all that of grain, will be taken away from the nature and elemental spirits. There will be heating installations, not only for winter gardens and smaller spaces for plants to grow, but for whole cornfields. Deprived of cosmic laws, grain will grow in every season, instead of only when it grows of its own accord – that is, when it grows through the working of the nature and elemental spirits.[8]

Is there some inner life of Mother Nature of a kind we may not at once be able to fathom? Some years ago a rather dramatic confirmation appeared of some words spoken by Rudolf Steiner in 1923. He warned that:

> Now you imagine that an ox suddenly decided that it was too tiresome to graze & nibble plants, that it would let another animal eat them and do the work for it, and then it would eat the animal. In other words, the ox would begin to eat meat, though it could produce the meat by itself.
>
> It has the inner forces to do so, what would happen if the ox were to eat meat directly instead of plants? If an experiment could be made in which a herd of oxen were suddenly fed with pigeons, it would produce a completely mad herd of oxen. That is what would happen. In spite of the gentleness of the pigeons, the oxen would go mad.'[9]

The experiment was made, not with pigeons but with ground-up remains of other ruminant animals fed to cattle. At huge expense, 'BSE' (mad cow disease) resulted in millions of animals being slaughtered.

---

[8] *The Karma of Untruthfulness* Vol. 1, (lecture given on 25 Dec, 1916 at Dornach, Switzerland) Vol. 1988, p.208.
[9] Lecture by Rudolf Steiner January 13, 1923, in Dornach, Switzerland, *Agriculture* lectures, Ch.2, ref. 37.

Dr Steiner made a key statement concerning the Full Moon, in his 'Agriculture' lectures:

> There is a definite connection between the Moon and the water in the Earth. Let us therefore assume that there have just been rainy days and that these are followed by a full Moon. In deed and in truth, with the forces that come from the Moon on days of the full Moon, something colossal is taking place on Earth. These forces spring up and shoot into all the growth of plants, but they are unable to do so unless rainy days have gone before... we have to raise this question: How should we best consider the rainfall and the full Moon in choosing the time to sow the seed? For in certain plants, what the full Moon has to do will thrive intensely after rainy days and will take place but feebly and sparingly after days of sunshine. Such things lay hidden in the old farmers' rules ...

The evidence we have here examined has greatly supported that bold claim.

For many years, graduation-day at Princeton University took place on a Sunday in June, outdoors, and everyone concerned hoped that the Sun would shine. There is in general more sunshine on Sundays, owing to less traffic and car exhaust and the reduction of air pollution. But, bizzarely, over the years, records seemed to show that the Sun shone more often on that particular Sunday of the year, in the neighborhood of Harvard University, than other Sundays. Were the clouds responding to the hopes of the academic community?[10]

That would be hard to credit. Or, could it be that we live in a world which does just occasionally take notice of our hopes and fears? We had better be careful what these are, if the world is liable to respond to them. They are not 'merely subjective' if they can make the Sun come out! Cleve Backster has republished his classic opus about what he calls 'primary perception' amongst plants, whereby they respond to the feelings of their owners, or how one plant can respond electrically if a neighbouring plant is chopped.[11]

Trees share their fears[12]: under attack, they are prone to secrete a mild poison into their leaves, causing indigestion to predators. They can even communicate their distress to others: if electrodes are attached to a tree-trunk, and a neighboring tree thirty meters away is struck with an axe, the

---

10 Roger Nelson, 'Wishing for Good weather, a Natural Experiment in Group Consciousness', *Jnl. for Sci. Exploration,* Spring 1997.
11 Cleve Backster, 'Primary Perception: Biocommunication with plants, living foods, and human cells', 2003; discussed in Tompkins and Bird *The Secret Life of Plants* 1973, also Lyell Watson, *Supernature* 1973, p.247.
12 Lyell Watson, *Supernature II* 1986, pp.64-67; Baldwin and Schultz, 'Rapid Changes in tree-leaf chemistry, Evidence for Communication between plants', *Science,* 1982, 221, pp.277-9.

Conclusion

former's trunk has been found to quiver with millivolt oscillations, by way of recording the attack.

Does a hormone signal released into the air cause nearby trees to defend themselves with tannin?[13] What the modern farmer may need most, even more than a new tractor or a Euro-grant, is a more vitalistic world-conception, whereby the alive-ness of the farm is experienced.

Through the twentieth century the inorganic, technical-scientific ideology continued to drive farming, spraying inorganic chemicals, mainly toxins, onto the fields and driving ever more farmers off the land. That is the end-result of a historical process, well-expressed by the historian Caroline Merchant:

> Between the sixteenth and seventeenth centuries the image of an organic cosmos with a living female earth at its centre gave way to a mechanistic world view in which Nature was reconstructed as dead and passive, to be dominated and controlled by humans.[14]

Prince Charles has made a comment upon Dr Steiner's 'Agriculture' lectures. In his new book 'Harmony,' Prince Charles had discussed the chemical-agriculture approach developed in Germany by Liebig in the 19th century:

> A century after Justus von Leibig conducted his pioneering research, the philosopher Rudolf Steiner was asked to deliver what became a famous set of lectures on the emerging crisis in agriculture. Steiner was quite clear about his view of this development. He described Liebig's approach as taking agriculture out of the realm of the living and putting it into the realm of death. Only in the realm of death, said Steiner, does a theory like this work...
>
> Despite Steiner's warning a century ago, together with two world wars and a dramatic acceleration in the world's population throughout the twentieth century, pressure has mounted on agriculture to adopt the clinical efficiency of the factory production process.[15]

Charles has been widely scoffed at for his belief in lunar-gardening calendars.

When Isaac Newton used the word 'menstrual', he solely alluded to an object in the sky, to a period which it measured out. It had no personal meaning for him, certainly no feminine meaning. Men would use the term,

---

13 For how trees respond to their leaves being eaten, see Lyell Watson *'Supernature II'*, 1986 pp.62-8; J. Narby, *Intelligence in Nature, an Inquiry into Knowledge*, 2005 USA.
14 Caroline Merchant, *The Death of Nature* 1990, p.xvi
15 HRH Prince Charles et. Al.,*Harmony a new way of looking at our world* 2010, pp 162, 165.

three centuries ago, concerning their equations. When we now use the term it solely alludes to the flow of blood and lacks any astronomical meaning. Women use the term concerning their experience. So the word has journeyed from one end of our world to the other, from outer to inner. We suggest that the word only has its proper significance from combining both these meanings; indeed we are in danger of forgetting what it means to be human if we fail to do this. The word alludes to a personal experience which in some way - however indirectly - has an astronomical timing. There is no clock, but there is a cycle in heaven, and doctors are keen that womankind should not understand this. We need a new kind of logic which apprehends this interconnection.

Our neighboring planet Mars has a couple of small moons in a close, low orbit, tumbling about like a couple of pock-marked potatoes, moving in opposite directions: how different from the sublime movements of Luna. She moves serenely in her own orbit-plane, neither that of the equator nor ecliptic (as do those of Mars), and she faces ever Earthwards. Thirteen times she revolves on her axis per Earth-year. I don't believe that today's astronomers have any even half-credible story of Her origin. They are presently on their fourth theory of Luna's origin – and shouldn't be allowed any more!

Selene's Sphere has been visited in our lifetime. Nine astronauts walked on her surface. Upon returning back home, they found it hard if not impossible to answer the question 'What did it feel like?' They could all recall clearly enough the operating instructions they had been given, but were strangely unable to find words about the experience itself. Daft theories sprang up, that the astronauts had not really been there - on account (I suggest) of the strange absence of credible personal narrative. Neil Armstrong, the first man to step on the Moon, became a hermit for years afterwards, and no journalist could get near him. He did however come out with a few cryptic words on the occasion of his giving an address at the twenty-fifth anniversary of the Apollo 11 landing in 1994, for which we should be grateful:

> There are unimaginable wonders [out there], for those who can remove some of truth's protective layers.[16]

It is a sphere of mystery. We've tried to remove a few of 'truth's protective layers,' and our struggle with these matters may help us to better understand how we belong in this world.

---

16 Neil Armstrong's address at a White House ceremony on July 20, 1994, on the 25th anniversary of man's landing on the Moon on July 20, 1969, started with him comparing himself to a parrot, 'saying only what he had been told to say': Richard Hoagland and Mike Bara, Dark Mission, the Secret History of NASA, CA 2007, pp.389 and 471.

Appendices

1. The 1940 Mather Moon-Phase Seedling Trials

The first thorough, scientific time-experiment with crop yields in relation to the lunar cycle, at least in the UK, was surely that of K. Mather, working at the John Innes Institution.[392],[393] It was in the year 1940, and some early results of Kolisko's had by then been published,[394] and although he does not mention her by name we may presume that it is her work that is being alluded to. He was testing the hypothesis that crops germinated better and grew better if sown just prior to the Full Moon, in fact two days before, which was Kolisko's argument. He published his results in the Journal of the Royal Horticultural Society in 1941, and his articles are the only ones ever published in that journal on the subject of lunar influence.

Mather sowed six rows of crops four times a month - two of tomatoes, four of maize - at each lunar quarter, or rather a couple of days prior to each lunar quarter, and he did this over 17 weeks, i.e he sowed 17 x 6 = 102 rows altogether. Each row was harvested after the same 33 days of growth. If indeed any lunar quarter was having some influence, it surely ought to show up in a so well-designed and carefully-executed experiment.

He analysed his data by taking means of pairs of rows (his sowings of tomato, Maize 1 and Maize 2) so that he ended up with three sets of data instead of six, and he then put a five-point moving average through each of these three and subtracted these out from the data. Thereby he removed the seasonal trends. He then grouped the several data-sets by lunar quarter to see if there was any significant difference.

Use of a five-point moving average has the effect of losing the first two and last two yield-values; so Mather's data ended up extending over three complete lunar months, after he had subtracted out his seasonal trend (he had one last row of a New Moon which he ignored and we'll follow him in this, it's easier if there are the same number of rows for each lunar quarter).

He obtained:

---

392 K. Mather & J. Newall, 'Seed Germination and the Moon' *Journal of the Royal Horticultural Society* 1941, 66, pp.358-66.
393 K. Mather, 'The Effect of Temperature and the Moon upon Seedling Growth' *Journal of the Royal Horticultural Society*, 1942, 67, pp.264-70. Kolisko wrote, 'Maize is a plant which needs the forces of the Full Moon at sowing time. Maize planted at New Moon does not do well' (*Moon and Plant Growth* 1936, p36); by which she meant, that sowing it two days before the Full Moon was optimal. It would appear that the Mather trials chose maize in accord with her recommendation.
394 E. and L. Kolisko, *Agriculture of Tomorrow*, 1940, Ch. 2 Moon and plant Growth.

Seedling weight after 33 days, group means & standard deviations
- 1st quarter   1.06 ± 13 oz. (n=9)
- Full Moon   1.45 ± 3.0 oz. (n=9)
- 3rd Quarter   -0.79 ± 2.3 oz. (n=9)
- New Moon   -0.11 ± 1.3 oz. (n=9)

His mean weights for each of his six boxes of seedlings, per lunar quarter, were around 3-4 ounces. Overall, his results gave, for Full Moon sowings 1.45±3.0 oz. (n=9) and the rest -0.26 ±1.77 oz. (n=27), which gives a t-value of 2.0 which is only very marginally significant, at say 1 in 20. He dismissed it, claimed to have obtained a negative result. It is better to first transform the six data-sets so that they have means of 100, because in order to merge separate data-sets they need to have the same arithmetic mean. Figure 1 shows the six data-sets, thus transformed. We then follow his example in putting five-point moving averages through the data-sets and grouping the trend-corrected data by lunar quarters, which gives us:

Seedling weight after 33 days, group means as percentiles
- 1st quarter   5.5 ± 14 oz. (n=18)
- Full Moon   11.2 ± 25 oz. (n=18)
- 3rd Quarter   -9.2 ± 36 oz. (n=18)
- New Moon   -7.1 ± 14 oz. (n=18)

Is this a significant result? There was overall a 18% yield difference between the Full and New Moon sowings (11.2+7.1 = 18.3, significant at t = 2.6), as predicted, at just over one lunar month after sowing. Note that we have kept the six groups separate, we did not merge them together in pairs as Mather did. The primary hypothesis here appears to be confirmed.

For a statistical test we may prefer to form just two groups, viz. those sown just prior to the Full Moon, and all the others. Merging these three other groups together gives us a mean of -2.7% ± 27 (n=54). This gives us a Full Moon excess of 11.2% (n=18), as we've already seen, and the others. That overall 14% yield increase is significant at a level of 1 in 50 (t=2.2). Mather has clearly obtained a significant result of large enough magnitude to be of relevance to the practical farmer.

Strictly speaking, we should use a four-point moving average rather than a five-point: if we are testing a hypothesis involving a fourfold pattern, as is the case with this data of sowings at lunar quarters, it is preferable for the moving average to be of this same length (or a multiple thereof, e.g. an 8-point moving average. Any other, e.g. a 5-point will tend to interact with and subtract out the effect in question). A four-point moving average is achieved by taking five points, but giving a half-weighting to the first and fifth value. Repeating the analysis in this manner gave:

Full Moon sowings 12.7±23 (n=18),

Others   -3.8 ±23 (n=54), t=2.6

That is a notably better result, indicating a 15.5% yield increase, and that improvement tends to give us confidence that a real effect is here present.

It would appear that some men with degrees and established positions wished to discredit the views of a German woman' while war was raging, and to do this they were prepared to be careless in interpreting their own very careful experiment. There is no doubt that Frau Kolisko was demoralised by such 'refutation' of her findings. But, objectively speaking, this Mather experiment must be accepted as the most thorough, independent vindication of the Moon-plant work of Kolisko, at least as published in the UK, in the 20th century.

The 1990 Tree-germination Zurcher field trials, in Rwanda

Over the six lunar months May-October 1990, four batches of the tree species *Maesopsis eminii* were sown fortnightly at the Full and New Moons in a tropical forest, with fifty grains per batch. Ernst Zurcher scored three main parameters: the days for the first germination to appear, the total seed germination, and the maximum height per batch after four months' growth. Overall, germination of the Full Moon batches was 56% higher than at New, and the maximal tree heights at the Full averaged 27% more than at New. In addition he counted the first day of emergence of the tree shoot and this was 23% more for the Full than the New, i.e New Moon shoots were slower to appear. These results are so remarkable that one would like to see the experiment replicated not using a lunar calendar, e.g. just sowing once a week.

Data summary: taken from Tables 1 and 2 of his 1992 article published in the Journal *Forestier Suisse*. Days for first appearance of tree-seedlings:

Full 47.3± 9.4 days (n=24), *New* 58.6± 19.2 days (n=23), t=2.5 significant at 1 in 100

For the number germinated (out of 50 seeds),

Full 14.5±7.9 (n=24), *New* 9.7±7.5 (n=24), t=2.1, just significant at 1 in 25.

For maximum height reached per batch after four months,

Full 20.1±4.9 cm. (n=6), *New* 15.8±6.7 cm. (n=6), t=1, not significant.

See Chapter 3 for Zurcher's work.

2. Excerpts from the BAH article (2001) by NK and Staudenmaier[395]
(Concerning data presented in Chapter 6)

There were, Steiner claimed, four types of 'formative force' that worked in the realm of nature. These, he said, did not work centrically as the forces known to modern physics, but rather worked via form, and he called these forces 'etheric' (Wachsmuth, 1932). This was a reformulation of the ancient doctrine of the four elements, conceiving them as process rather than as substance.

---

395 'Mond-Trigon-Wirkung: eine statistiche Auswertung', Lebendige Erde, November 1998, pp.478-483.

Within the Anthroposophical movement, botanical studies of plant morphology by Bockemühl have supported the view that the stages of plant growth may be seen in terms of such 'formative forces' that are linked with the traditional four elements. He has related the stages of leaf, flower and seed formation with water, air and warmth (Bockemühl, 1985). But, especially in Germany, Biodynamic experts had expressed unease about using these sidereal element-rhythms in the calendars, viewing them as inappropriate and lacking endorsement from Steiner (Koepf et al., 1996).

Ulf Abele

Dr Ulf Abele analyzed his 1980 potato data (see Chapter 4) by subtracting from each set of twelve weight yields the corresponding values on its regression line, then separating the difference values into two groups: the three with predicted yield maxima, and the remaining nine. The results are shown in Table

Four years trials of the 'Thun effect' by Abele.

Crop	Data	Mean Y	Trigon predicted (t ha^{-1})	Other rows (t ha^{-1})	t-value
Mean yields and yield deviations					
Barley	1970	3.67	0.29 ± 0.15	−0.10 ± 0.21	2.6 > t_{10}, 0.05
Oats	1971	5.46	0.22 ± 0.45	−0.07 ± 0.15	1.6 > t_{10}, 0.10
Carrot	1972	9.62	1.29 ± 0.60	−0.43 ± 0.61	3.9 > t_{10}, 0.005
Radish	1973	26.25	4.81 ± 1.05	−1.61 ± 2.10	4.7 > t_{10}, 0.001
Combined data with yield deviations normalized to a mean of 100					
Barley + oats	1970, '71		5.7 ± 6.5 (n = 6)	−1.9 ± 4.4 (n = 18)	3.1 > t_{22}, 0.01
Carrot + radish	1972, '74		15.9 ± 5.8 (n = 6)	−5.3 ± 7.3 (n = 18)	6.2 > t_{22}, 10^{-4}

Table: Ulf Abele sowing trials 1970-73

Each of Abele's four trials showed yield maxima where predicted, and yield deviations from the seasonal trend are positive for the trigons predicted and negative for others. The differences of these means are statistically significant except for the 1971 trial. Significance was investigated by a t-test, whereby the hypothesis of zero difference between the group means was rejected. For Abele's trials, the group means differ with a high probability of 95% - 99%.

It is preferable to use groups of more than three for statistical treatment, as a t-test requires that the two groups being compared are normally distributed, and one cannot ascertain this from merely three values. These trials were therefore combined with those for grain yields (1970 and 1971) and those for root crop yields (1972 and 1974). To do this, the yield-values were first transformed to the same mean of 100, enabling the data-sets to be combined. They were then corrected for seasonal-trends by Table 1 subtracting out the linear regression lines as before. These combined sets gave the yield data as shown in Table 1.

Statistically, the grain yield excess was significant at $p = 0.01$, while the root-crop yield over Abele's two trials was significant at $p = 0.0001$. Expressing the mean yield excess in the predicted trigon as a percentage for Abele's two grain crop trials, the 'fruit-day' trigons showed a mean yield excess of 7% over the others (5.7+1.9 %), while his two root crop trials averaged an excess of 21% (15.9+5.3%), from sowings in the predicted trigon as compared with sowings at other times. Spiess viewed these results as positive (Spiess 1994).

Ursula Graf

Three years of potato trials by Graf for her doctorate thesis at the Zurich ETH, investigating the sidereal trigons, obtained equivocal results. Graf & Keller commented that Graf's 1975 growth chamber experiment with radishes on Biodynamically treated soil gave a yield excess of 20%, significant at $p < 0.001$, for yields from root-day sowings as compared with the others, for a total of 264 radish grown (Graf & Keller, 1979).

A review of the evidence by Dubrov alluded to "very strict standardised experiments ..." as having shown how lunar influence worked in plant growth and development (Dubrov, 1996). He referred to Kolisko, but also to the Thun-Heinze and Graf-Keller reports, both published in 1979.

The Radish trials of Hartmut Spiess, 1982

In the 1990s, discussions in print of the Biodynamic calendar in Europe, America and New Zealand, have alluded to the experiments conducted by Spiess as having tested the Thun-hypothesis and failed to replicate it (e.g., N.Z. Biodynamic Association 1989; Llewellyn, 1993). Enjoying widespread publicity, and published by the Forschungsring of the German Biodynamic movement, the Spiess results have worked to discredit Biodynamic calendars.

We suggest, however, that although the experiments were well designed, this was not matched by a corresponding care in the data analysis. While the Spiess sowing trials and methodology had commendable features, their use of parabolic curves to model the seasonal trend was inappropriate, and the alternative here advocated of moving averages to model the seasonal trend gave element-means of considerably smaller standard deviations and thereby found the significant results.

As well as his 1982 radish trial, there were two others he performed in 1979 and 1980 over 30-day periods. In the first case yields increased by a factor of seven from start to finish of the experiment, and in the second case they more than doubled. But in experiments conducted over only one month, containing such large seasonal trends within the data, it is unrealistic to expect low-amplitude sidereal rhythms to be detectable

To compare the yields (Y) of two or more trials, normalized yield values $Y1 = 100Y/Ymean$ were used. For investigating whether lunar-monthly rhythms are present in crop yield data, it is in general necessary to allow for the seasonal or year trend, as this may be larger in magnitude than rhythms of a monthly nature. To do this, one can put second-order regression lines through the data, i.e. best-fit parabolae, as did Spiess, or one can apply a moving average.

Such a moving average here includes two sowings before a given date and two after, taking the mean of these five values. The latter method was used here (Kollerstrom & Staudenmaier, 2001). Both techniques, a regression line, plus a five-point moving average, are illustrated in the Figure, applied to the Y1 data of the 1982 radish experiment.

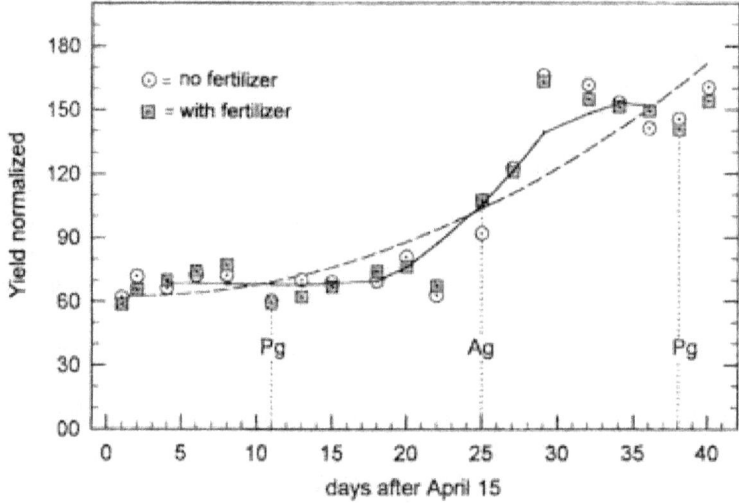

Fig. 11.1 Spiess radish trial, 1982

The trend line values (T) are then subtracted out from the Y1 data-points. That subtraction gives a new data-set, which is called Y2 (i.e., Y2=Y1-T). These Y2 data points are used to test for lunar-sidereal rhythms present in final crop yield.

We used this methodology as applied by Spiess, but modeling the seasonal trend differently as described, to go from Y1 to Y2. As the Figure indicates, the way one allows for seasonal trend can affect final values considerably.

The 1982 experiment involved one lot of sowings per Moon-constellation (i.e. 12 per sidereal month) over a 39-day period, so that 19 rows were sown in all. Two separate sets were conducted in parallel, one with added fertilizer and one without. The experiment had 'randomized, complete block designs with four replications' at each sowing (Koepf et al., 1996).

The next step involved making a separation within the Y2 data-set. The hypothesis tested was that one 'trigon', in this case root-day sowings, gives a better yield than do the others. The data was therefore separated into two groups, root-day sowings and the rest. Spiess separated the data into four groups, by the four trigons, performing a four-way analysis of variance.

These two approaches are both valid, but the former may have simpler statistics. Separating the Y2 yield values into these two groups gave the results shown in Table 2. These Y2 mean values are percentages, because the Y1 groups have means of 100.

Thus the no-fertilizer group had 7.6% excess for the root-day yields, while the other rows had 2.4% deficit, so that, overall, the root-day sowings achieved 10% more in weight yield than did the other sowings. Owing to the large scatter in the data the effect was not statistically significant ($t=1.7$, $p = 0.1$). It is preferable that mean yields should not increase so sharply during the course of an experiment. Even with a 10% excess as predicted, this increase prevented the data from attaining significance. Trial II with fertilizer did not show a statistically relevant yield distribution.

Thun has claimed that the sidereal trigons show up best if crops are not treated with fertilizer (Thun 1964), which Graf in her PhD at Zurich claimed to have

TABLE 2

Mean yield-deviations (Y_2) normalized to a mean of 100 for Spiess's two 1982 radish trials, root-trigon sowings vs. others.

	Root-days	Other rows	t-value
Trial I (no fertilizer)	7.6 ± 17 (n = 3)	−2.4 ± 10 (n = 12)	$1.7 > t_{13}$, 0.1
Trial II (fertilizer)	4.7 ± 17 (n = 3)	−0.4 ± 7 (n = 12)	$0.8 > t_{13}$, 0.5

confirmed (Graf 1977). Though of smaller magnitude than others have found earlier from investigations of the 'Thun effect', this yield excess may nonetheless be large enough to be of interest to commercial growers, as the result of a well-designed and carefully performed test of the 'Thun-effect'.

Carrot trials, 1978-80

This same technique of data analysis was applied to the three years of carrot sowing data, which Spiess performed in 1978-80 (Spiess, 1993, 1994). For these three years of trials the same experimental procedure was used, sowing approximately once per Moon-zodiac constellation as for his 1982 radish trial (see previous section). The trials extended over a month, which meant that 14-

TABLE 3

Standard deviations in the Y_1 and Y_2 data groups of Spiess trials here analysed.

Crop	Date	Y_1 group	Y_2 group
Carrot	1978	100 ± 11.8 (n = 14)	−0.3 ± 2.8 (n = 10)
Carrot	1979	100 ± 8.9 (n = 10)	0.4 ± 6.8 (n = 6)
Carrot	1980	100 ± 10.7 (n = 15)	−0.3 ± 4.3 (n = 11)
Radish	1982	100 ± 40.0 (n = 38)	0.1 ± 10.2 (n = 30)

15 rows were sown each year. However, the last six rows sown in 1979 suffered from a two-week discontinuity in the data so had to be omitted. The carrot trials achieved a more uniform growth than did Spiess's radish trials. An indicator of how well trials have been conducted is the standard deviation of the yields expressed as a percentage of the mean. This is shown in Table 3, above, where the variance of the carrot-yields is compared with that of the radish-trial discussed above. For whatever reason, these are a mere one-third the value for the carrot trials, so that better results might be expected from analyzing his carrot-data. Five-point 'moving average' trend lines were put through these groups as before, and subtracting these gave the Y2 values. The Table shows how two-thirds of the variance is removed by subtracting out the seasonal trend line, passing from Y1 to Y2, also how the number of rows per group was reduced in passing from Y1 to Y2 values, by use of the moving average.

The transformed carrot data 1978-80 were separated as before into two groups, root-day sowings and others. Over the three years of these trials, the Y2 root-day values averaged 6.3 ±4.1 (n=5) while other sowings averaged -1.6°± 2.9 (n=22), an 8% excess, with a t-value of 5.4, which was highly significant (p = 0.001). (2)

No Perigee effect?

Spiess averred that his experiments found yield increases at perigee (nearest distance to Earth). He drew this as the *first conclusion* from his years of lunar-horticulture experiments. Thus his summary of conclusions began:

> An influence of the anomalistic rhythm (perigee, apogee) existed for all cultivated crops. Plants sown at Moon's perigee all showed positive reactions.

The groups of data here analyzed, of final crop yield plotted by lunar distance at sowing date, do not show any such result. Two graphs are here shown, of his carrot data which has a few % increase in yield at perigee, and then his radish data which had a slight decrease in yield. Overall, one may query whether his data shows this result as he claimed

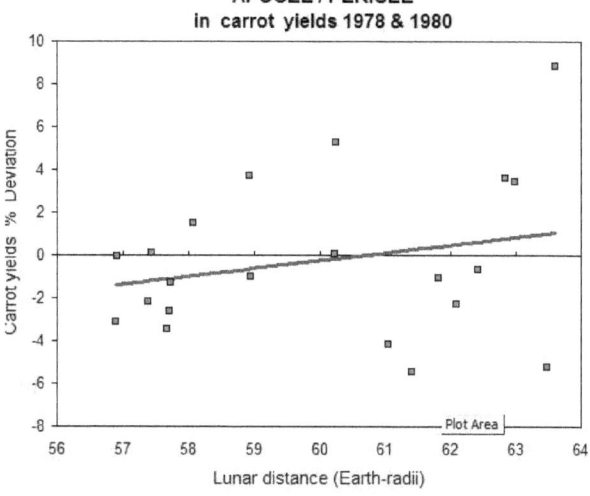

Fig. 11.2,3 Spiess carrot and radish yields, plotted by lunar distance at sowing

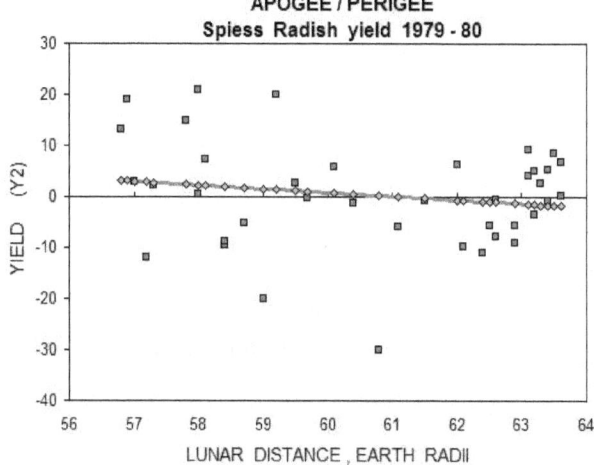

3. Star-rhythm in Mistletoe Shape

Tree-buds respond in their shape, to a fortnightly lunar rhythm (Chapter 8). This small-amplitude change is only one or two percent. The 'lambda-function' (as Lawrence Edwards called it) is a highly inscrutable term derived from projective-geometry, which express eg how 'egg-shaped' an egg is. Two Swiss investigators looked at this with mistletoe berries. With data over the years 1995, 1997, 1998, 2000 and 2001, they photographed each bud as required and took the lambda-measurements, but could find no such rhythmic shape-change.

Biology professor John T. Burns was visiting the Hiscia Klinik in Arlesheim, where this experiment was being conducted. As a cancer clinic, it prepares medicine from the mistletoe berry according to indications given by Dr Rudolf

Steiner. Burns, from Bethany College, West Virginia, had earlier published the very comprehensive review, *Cosmic Influences on Humans, animals and Plants, an annotated bibliography* which examined hundreds of different kinds of solar and lunar biological rhythms (an opus highly relevant to the present treatise, for its one-paragraph summaries covered many of the investigations here alluded to, up until 1997) John Burns scrutinised their data, and pointed out to them, that they had a sidereal rhythm present, cycling three times per sidereal lunar month. In other words, they had a Thun-type effect of nine day periodicity - instead of the fourteen days they had expected, which Lawrence Edwards had found!

Their mistletoe seemed to be responding to the round zodiac of the stars. Was this another example, of what Kolisko in 1927 called 'The working of the stars in earthly substance'? Stefan Baumgartner was employed by the Hiskia Klinic to ascertain whether the mistletoe medicine 'worked'; i.e., whether it cured cancer. That large question is beyond our scope here, but he and Heidi Fluckinger did conduct a very careful sequence of experiments.

Earlier, they had reported on how the changing shape of the mistletoe berry as it grew and ripened fitted a 'path curve' surface, which could be described by the lambda-function. Mistletoe berry lambda-values vary from about 0.8 to 1.2, 'which corresponds to downwards or upwards pointed egg-shapes respectively.'[396],[397] As the berries ripened from June to December, they would change from slightly upwards-pointed ($\lambda>1$) to slightly downwards-pointed ($\lambda<1$). A trend line for each year was subtracted out from the data. I published a review of their work, which accepted the way they had done this.[398] Modulating this trend line, was a nine or ten day rhythm in shape, and was that linked to the moon in the zodiac? What they called the λ' value, were the residual lambda-values, after these trend lines had been subtracted out.

If the rhythm were sidereal, that would conveniently mean that all five years of the data could be added together. Each data-point has its own sidereal lunar celestial longitude, when the berry was photographed, and these can be plotted on the same graph: that can't be done with the tree-bud data-sets, because the phase and length of the lambda-waveform are not exactly known, so each year's data has to be plotted separately.

The two investigators plotted their transformed λ' data by the twelve unequal constellation-divisions (Figure 1 – from the authors' 2003 paper, here reproduced with kind permission). In the middle one can see the long Virgo

---

396 Stephan Baumgartner, Heidi Flückiger and Hartmut Ramm, 2002, Elemente der Naturwissenchaft, 'Form...Mistelberen', 77 (2) pp. 2-15. This was followed by their 2003 article, 79 (2) pp.2-21.
397 Baumgartner et. al., 'Shape changes of ripening mistletoe berries', Archetype No. 9 (Ed. David Heaf), September 2003; followed by, 'Mistletoe berry shapes and the zodiac,' Archetype No. 10, September 2004, (This journal is now defunct.), p.1.
398 N.K., 'Star-Rhythm in Mistletoe Shape' Elemente der Naturwissenchaft 2011, 94, 119-124.

constellation, and next to it the very short Libra. The horizontal line along the middle of their graph shows the Moon's path every 27 days. Their data suggested a peak in the Water-element, closely followed by Air.

The authors were not prepared to part with their data but allowed me a high-resolution graph, inviting me to extract lunar longitudes and mean deviations of their lambda values therefrom. I did this and have used these; an error of a degree or so is likely as resulting from my data extraction.

The authors did not find any physical or environmental factor correlating with the bud shape, eg rainfall or humidity, maybe because of the nearly circular shape of the buds, whereby the lambda-value hovered around unity: ambient moisture would have expanded or contracted the berries, without altering their lambda-value (the 'path-curve' shape).

Using a 3rd-Harmonic Waveform

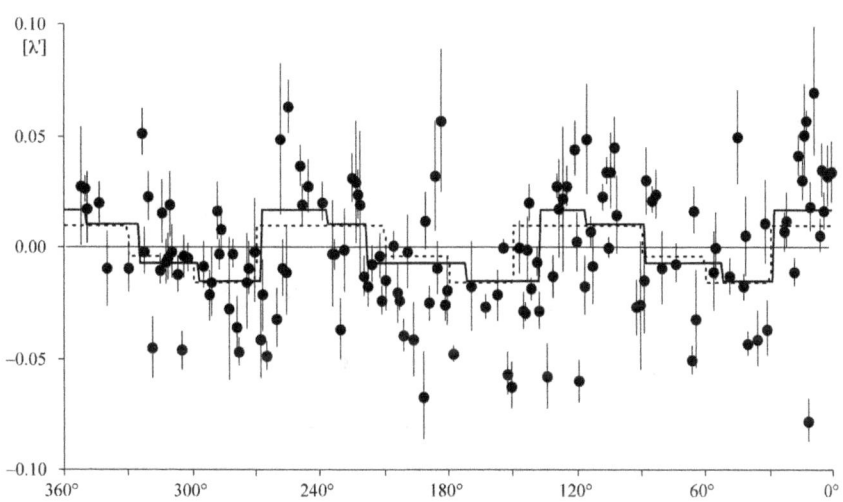

Fig 11.3: From the Authors' 2003 paper. Total (139) trend-corrected λ-values from mistletoe berries over six years, plotted by (tropical) lunar longitude at measuring-time, and showing the 12 unequal-constellation boundaries used by the authors.

The trend-corrected data here used is zero-sum, i.e. its average is zero

I plotted these 139 combined λ' data points by lunar zodiac longitude at time of measurement, *using the star-zodiac*. The plot extended to 120°, through four zodiac signs, so that longitudes of 130° or 250° from zero Aries would count as 10°. The best-fit waveform for this data is shown in Figure 2. Plotting such a 120° 'third harmonic' waveform assumes that the sidereal zodiac elements are divided in a twelvefold manner (by the 'trigons') *at equal 30° intervals*.

The data indicated that the Four Elements – or, four ethers – work into berry (not bud) morphology. The plot shows a sort of Thun-type 'sidereal element' effect, but spread over both Air and Water ('flower-days' and 'leaf-days'). I've subtracted 25° from their given Moon-zodiac longitudes (i.e., normal 'tropical' longitudes, as calculated by the Authors) to give sidereal-zodiac longitudes: that being the generally-agreed current 'ayanamsa' or phase-difference between tropical and sidereal zodiacs.

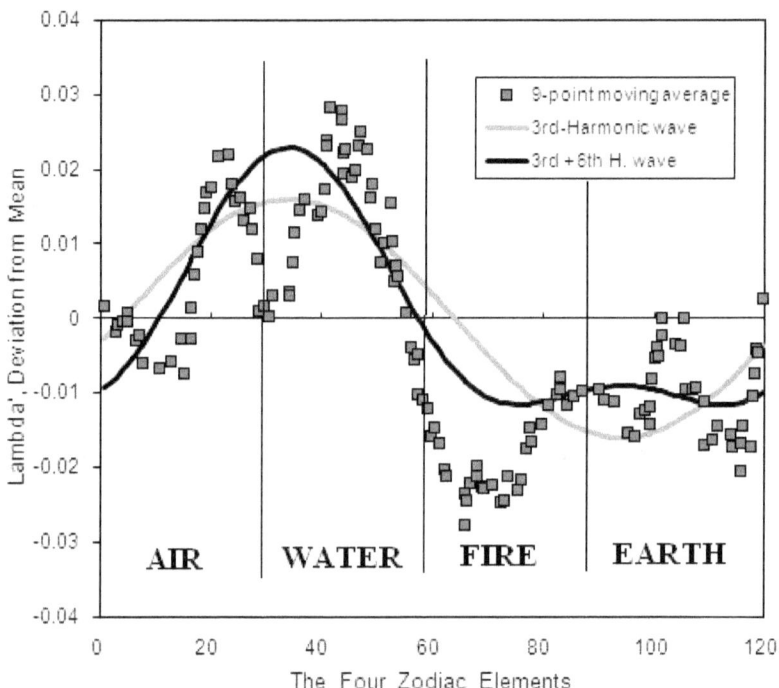

Fig. 11.4: Two best-fit waveforms put through six years of trend-corrected mistletoe lambda-values, taken data directly from fig 1, with lunar celestial longitude using the star-zodiac.

A 3rd-harmonic waveform will have the equation, $\lambda = \alpha \sin 3(l - \beta)$ where α is the *amplitude* of the effect, the number '3' defines the *wavelength* or frequency, so the waveform goes through three cycles per sidereal lunar month, measured 0 - 120°, and 'β' gives the *phase* of the waveform. The variable in this equation is 'l,' the lunar zodiac longitude. The equation gives us the expected lambda-value.

The waveforms are a 'best-fit' against the 139 data-points: that is to say, their parameters of amplitude and phase were adjusted to minimise the sum of the squares of the vertical y-distances between the sinewave and the data-points.

I've put in an extra waveform at half the wavelength (ie, a 6th-harmonic), having the same peak in celestial longitude as the primary third-harmonic waveform:

$$\lambda = 0.016\sin3(l - 4°) + 0.007 \sin6(l-19°)$$

As well as these two sinewaves, I've plotted a moving average (13-point) through all the given lambda-values. The actual data points (shown in Fig 1) would have a much larger scatter.

Amplitude of the Effect

How big is the effect? The authors claim that a 4% difference in lambda-values may be found between two halves of the data. The graph function shows peak amplitude in the lambda function of 0.023, which we may express as ± 2.3%, and this tends to confirm the Authors' view. Or, from inspection of Figure 2, one can generally confirm the Authors' view of the amplitude of the effect which they discovered.

This is a slightly larger fluctuation than Lawrence Edwards found for his fortnightly bud-rhythms. I found amplitudes of 1.3% and 1.8% lambda in cherry bud rhythm and beech bud rhythm. I suggest that the wave-harmonic approach, gives a more straightforward means of describing the amplitude of the effect.

Significance level

The 1995 data-set was the one which the two authors used to formulate their hypothesis, namely that λ-values increased for Moon in Air and Water signs and decreased for Fire and Earth signs. That year of data therefore needs to be excluded from any statistical test. Using Figure 2, their hypothesis would compare the data spread over 0-60° with that spread over 60 -120°. This approach is simpler, but also more scientific and credible: no statistician is ever going to want to group the data by twelve irregular constellation-divisions. I'm unable to subtract out the 1995 data, but here is what a t-test gave, for six years of Mistletoe λ' grouped by lunar-zodiac element:

Air & Water: 9.4±29, (n=73) ; Fire & Earth: -14.4±25 (n=64), t = 5.1

(scaling up λ by 1000 to remove decimals) That t-value corresponds to around one in ten thousand. It would be a bit lower if we subtracted out the 1995 data. That is a remarkably high value.

Getting a decent significance level in biological data is notoriously difficult. So the Authors are to be congratulated in achieving this. That high level of significance is related to the very sharp boundary evident in Figure 2, at the 60° threshold where Fire/Earth changes over into Air/Water. The moving-average shows a sharper 'boundary' than the sinewave: if a further set of data were to confirm that, then one could add on one (or more) extra waveforms to model the sharper boundary. Such a sharp boundary at 60° tends to confirm that an equal-interval zodiac is here operating, as such a dramatic effect could hardly be shown with unequal constellations.

Relevance to Healing

The theory at stake here is intriguing, that mistletoe, a plant with no roots in the Earth, is more 'astral' than it is etheric in its life-energy, and so is attuned to a star-rhythm rather than the more normal 14-day tidal rhythm. It could be relevant, that the Water and Air elements are 'positive,' as indicating the balanced, etheric energies required for healing cancer, as applied in the Hiscia Klinic.

The Authors have made a contribution towards what Lili Kolisko called in 1927 'The Working of the Stars in Earthly substance,' the first step of which was taken by Maria Thun in 1956. Their finding also provides indirect support for the 'Thun effect' used in Bio-dynamic calendars.

Some might wish to take the view that the discrete-boundary approach (of unequal constellations) and the wave-harmonic approach here advocated are compatible, and are two different approaches to the subject, maybe like the wave/particle duality of modern physics. The authors utilised the unequal-sign constellations, 13 of which exist around the ecliptic according to the modern astronomical boundaries. But, a bio-dynamic textbook has now emerged,[399] which advocates, based somewhat on my research,[400] the ancient star-zodiac, as the proper framework for studying time-patterns of sidereal influence upon plant growth.

4. Solar-lunar sinewaves

To continue from Chapter 10, page 206: to construct a 3rd harmonic sinewave that models the sidereal lunar pattern. Adding on 45° to the lunar zodiac longitude will cause it to peak in Air (i.e., the trigon for 'flower days'). Or, if the equation needs to use the 'tropical' longitude rather than sidereal, because programs normally give that, then subtracting 24° or 25° converts from a tropical to a sidereal reference. So an 'ideal' Thun equation peaking in Earth will be:

$$Y/Y_m = 1 + A\sin3(M - 24° - 15°)$$

Where A is the amplitude i.e. how big the effect is. In writing this sine-wave-equation we are assuming that the yield-decrease in the Water trigon is as large as the Earth-trigon increase.

The Spiess data for three years of carrot sowings gave a low amplitude sidereal waveform, peaking in Earth:

$$Y/Y_m = 1 + 0.03\sin3(M - 24° -9°) \qquad \text{(Fig 4.15)}$$

Consulting that diagram in Chapter 4, we see how its 'trigon' peak of the waveform was displaced by 5 or 6 degrees, it wasn't quite centered in the 'Earth' trigon.

---

[399] 'Cosmos, Earth and Nutrition, The Biodynamic Approach to Agriculture' by Richard Thornton Smith, Sophia Books 2009: pp.150-151 for use of equal-interval zodiac.
[400] Robert Powell The Astrological Revolution (Lindisfarne books, 2010) has an Appendix 4 'Planting by the Moon,' which gives my argument for an equal-interval zodiac framework for use with growing crops.

The Bishop radish data of 1978 of Chapter 6 showed a large amplitude of 31%, and a sidereal-element ('trigon') peak in Earth displaced by eight degrees in the other direction, to:

$$Y/Y_m = 1 + 0.31\sin 3(M - 48°) \quad \text{(Fig 6.5)}$$

Analyzing the Bishop 1978 data by lunar day, as in this Chapter (Figure 10.1), we saw mainly two harmonics combining to make up that lovely pattern. Here the angle used is (L-D) where D is the horizon against which the Moon descends (Or strictly it's the zodiac longitude of that local horizon). The amplitude is rather large, so let us recall that no other data-set has as yet confirmed it. In our standard format this becomes:

$$Y/Y_m = 1 + 0.01\cos(M-D) - 0.31\cos 3(M-D) + 0.25\cos 4(M-D) \quad \text{(Fig 10.2)}$$

D is the descendant i.e. horizon, so the angle (M-D) rotates once per 24.8 hour lunar day. We use the Descendant rather than Ascendant in order to place the maximal effect position viz. moonrise at the center of the graph. At moonrise (M-D) = 180° when each of these three terms has its maximum value: they add up to 36%, which is huge. The first harmonic is almost absent, in this diurnal rhythm.

How closely are the lunar-day terms synchronized to moonrise? Adjusting the variables to (M-D-p), so they move coherently as 'p' is varied, the best-fit position turns out to be not more than two degrees away from that centered on moonrise. It's a fairly exact response to the Moon rising! The horizon moves 15° every hour on average, which implies that an influence is operative within minutes of sowing the seed.

This in turn suggests an astrological notion of causality, whereby some kind of 'imprinting' takes place at the moment of sowing, or at the first moment of water absorption, affecting future crop yield: just as Ptolemy said.

Geomagnetism in the Bishop data appeared as a linear affect, which we may represent as:

$$Y/Y_m = 239 + 5(G-23) \text{ oz.} \times 100 \text{ per radish} \quad \text{(Fig 10.2)}$$

where G is the geomagnetic field strength using the 'K' index. This implies a variation in crop yield of ± 28%, which is large. Such a correlation with the GMF is not always present in crop yield data.

We combine these three equations as they appeared in the Bishop radish trial, as:

$$Y/Y_o = 100 + 31\sin 3(M - 47°) - 10\cos 3(M-D) + 25\cos 4(M-D) + 2.1(G-23)\%$$

where M is lunar zodiac longitude, D is the descendant (earth's horizon) and G is the Earth's geomagnetic field strength, measured using the 'K' index, which ranged from 9 to 37 over this sowing period.

All three of these factors – sidereal, diurnal and geomagnetic - are of large enough magnitude to be of urgent importance to the busy farmer. That is the beauty of this crop yield equation. No solar terms are present in it; but, we bear in mind here that the mean 27.3- day period of solar rotation imprints itself into the GMF, giving an effect which is periodic but not cyclic.

Bibliography

Ulf Abele, PhD thesis, 'Vergleichende Untersuchungen zum konventionellen und biologischdynamischen Pflanzenbau unter besonderer Berucksichtigung vom Saatzeit und Entitaten', University of Giessen, West Germany, 1973.

Ulf Abele, 'Saatzeitversuch zu Radies', Lebendige Erde, 1975,6, 223-225.

Giorgio Abrami, 'Correlations between lunar phases and Rhythmicities in Plant Growth under Field Conditions,' Canadian Jnl. Botany, 1972, 50, 2157-2166.

Giogio Abrami and G. Piccardi, 'Seed Germination as a biological test for the study of fluctuating phenomena', Jnl. Interdisciplinary Cycles Research, 1973, 4, 267-82.

E. Adderley & E.Brown, 'Lunar Component in Precipitation data' Science, 1962, 137 749-750.

E. Andrews, 'Moon talk, the Cyclic Periodicity of Postoperative Haemmorage' Journal of the Florida Medical Association, May 1960, 1362-66.

'The Astrologer' (a quarterly), Foulsham's, London 1887- , gardening section.

Francis Bacon, Sylva Sylvarum (1672), in: 'The Works of Francis Bacon', Eds Spedding and Ellis, London 1887, Vol. 2, 636.

Michael Baigent, 'From the Omens of Babylon' 1994.

M. Baker, 'Gardener's Magic and folklore', Universe Books, New York 1978.

Adele E. Barger, 'Gardening Success with Lunar Aspects', American Federation of Astrologers, 1977.

Stephan Baumgartner, Heidi Flückiger and Hartmut Ramm, 'Form...Mistelberen,' 2002, Elemente der Naturwissenchaft, 77 (2) pp. 2-15; followed by their 2003 article, 79 (2) pp.2-21. English translations: 'Shape changes of ripening mistletoe berries', Archetype No. 9 (Ed. David Heaf), September 2003; 'Mistletoe berry shapes and the zodiac,' No. 10, September 2004.

B. Bell and R. Defouw, 'Concerning a lunar Modulation of Geomagnetic Activity' Jnl. Geophysical Research, 1964, 69, 3169-3174.

- 'Dependence of the lunar modulation of geomagnetic activity on the celestial latitude of the Moon', Journal of Geophysical Research, 1966, 71.3, 951-7.

C.F.C. Beeson, 'The Moon and Plant Growth', Nature, 1946, Vol. 158, 572-3.

- 'Forestry, Horticulture and the Moon' Forestry Abstracts, 1946, 8, p.191.

Simon Best, 'Lunar influence in Plant Growth: a review of the evidence', Phenomena, (Toronto, Canada) August, 1978, 23-24.

- 'The Moon and Plant Growth - A Review' The Astrological Journal, Spring 1978, 67-73.

- and N. Kollerstrom, 'Planting by the Moon 1980/81', Foulsham's 1980, then Astro Computing Services, San Diego US 1982.

'Biodynamics, New Directions for Farming and Gardening in New Zealand', New Zealand Biodynamic Association, Random Century New Zealand, 1989.

Colin Bishop, 'Moon influence in Lettuce Growth', Astrological Journal, Vol. 10.1, winter 1977/8.

Jochen Bockemühl, 'Elements and Ethers: modes of observing the world', in Towards a Phenomenology of the Etheric World, Anthroposophic Press, (J. Bockemühl,ed.) New York 1977, 1985, pp.1-68.

Donald Bradley, Woodbury and Brier, 'Lunar Synodical Period and widespread Precipitation', Science, 137, 1962, 748-9

Bradley and Brier 'Lunar Synodic Precipitation in the United States' Journal of the Atmospheric Sciences, 1964, 21, 386-395.

Frank Brown, 'The Rhythmic Nature of animals and Plants' Cycles, April 1960, 81-92.

- and Carol Chow, 'Lunar-Correlated Variations in Water Uptake by Bean Seeds' Biological Bulletin, October 1973, 145, 265-278.

- 'Biological Clocks: Endogenous Cycles synchronized by subtle geophysical rhythms,' Biosystems (Amsterdam), 1976, 8, p.75.

E. Bunning 'Endogenous Rhythms in Plants', Annual Review of Plant Physiology, 1956, 7, p.86.

John T. Burns, 'Cycles in Humans and Nature, an Annotated Bibliography' New Jersey, 1994.

- 'Cosmic Influences on Humans, Animals and Plants, an Annotated Bibliography' New Jersey 1997.

Harold Saxton Burr, 'Diurnal Potentials in the Maple Tree' Yale Journal of Biology and Medicine 1945, 17, 727-734.

W.J. Burroughs, 'Weather Cycles: Real or Imaginary?' CUP 1992.

H. Caspers, 'Rhythmische Erscheinungen ... und das Problem der lunaren Periodizitat bei Organismen' Archives Hydrobiol. Suppl., 1951 18, 415-494.

J. Cloudsley-Thompson, 'Biological Clocks, their Functions in Nature' 1980: Ch.7, 'The Moon and Life.'

T.Criss and J.Marcum, 'A Lunar Effect on Fertility' Jnl. of Social Biology, 1981, 28, 75-80.

Nicholas Culpeper, 'Culpeper's Complete Herbal, a comprehensive description of nearly all herbs, with their medicinal properties' Foulsham's, London.

Bruce Cumming, 'Correlations between periodicities in germination of Chenopodiun botrys and variations in solar radio flux', Candian J. Botany, 1967, 45, 1105-1113.

- 'Biological Cyclicity in Relation to some astronomical parameters: A Review' in Geo-cosmic Relations; the earth and its macro-environment, 1990, Amsterdam Ed. G. Tomassen, pp.31-55.

Robert Currie, 'Evidence for 18.6-year Signal in Temperature and Drought Conditions in North America since AD 1800' J.Geophysical Reserch, 1981, 86, 11055.

- 'Evidence for 18.6-year lunar nodal drought in Western North America during the past millenium,' J.Geophysical Res., 1984, 89, 1295-1308.

- 'Examples of 18.6-year and cyclic 11-year terms in World Weather Records' in: 'Climate: History, Periodicity and Predictability' Ed Rampino et al., NY 1987.

Winnifred Cutler, 'Lunar and Menstrual Phase Locking', American Journal of Obstetrics and Gynecology, 1980, 137, p. 834.

W. Cutler et al., 'Lunar Influences on the reproductive cycle in women', Human Biology, 1987, 59, 959-72.

'Dariotus redivivus, or a brief introduction conducing to the judgement of the stars', London 1653 The agricultural section has been inserted by a Nathaniel Spark into this edition.

Norman Davidson, 'Astronomy and the Imagination' 1985.

Geoffrey Dean and Arthur Mather, 'Recent Advances in Natal Astrology, a critical Review 1900-1976' Analogic Perth, 1977, pp.60-66.

Marie Delclos, 'Astrologie Racines secrètes et sacrées,' Paris 1994 (sidereal zodiac).

John Dewey, 'The Moon as a Cause of Cycles', Cycles, Pittsburgh, 1959, p.197.

- 'Cycles - The Mysterious Forces that Trigger Events', Manor Books, NY, 1973.

Alexandre Dubrov, 'The Geomagnetic Field and Life', trans. from the Russian, 1978.

- 'Human Biorhythms and the Moon' Nova Science Publishers, New York 1996.

Lawrence Edwards, 'The Field of Form, research concerning the outer world of living form and the inner world of geometrical imagination', 1982.

- 'The Vortex of Life, Nature's Patterns in Space and Time,' 1993.

Klaus Endres and Wolfgang Schad, 'Moon Rhythms in Nature' Floris 2001, from German, Biologie des Mondes, Mondperiodik und Lebensrhythmen, Stuttgart 1997.

K.J. Farbridge, 'Lunar Periodicity of Growth Cycles in Rainbow Trout,' Journal Interdisciplinary Cycles Research, 18 (1987), pp.169-177.

Kendrick Frazier, 'Our Turbulent Sun' Prentice-Hall, New Jersey 1987: Ch.11 on 27-day cycle.

J. Fritz, 'Studies on the influence of the synodic moon rhythm on the growth of Radish ...', 1994, a diploma in German at Kassel University, Wistenhausen (Cited by Zurcher).

Agnes Fyfe, 'Moon and Plant: Capillary Dynamic Studies,' 1967 Arlesheim, Switz. (Die Signature des Mondes im Pflanzenreich), 1975 2nd edition.

- 'The signature of the planet Mercury in plants, Capillary dynamic studies,' The British Homoeopathic Journal: Part I, 1973 201-232, Part II January 1974, pp.26-60, part III 1974, 111-124 ('Die Signatur Merkurs im Pflanzenreich, Kapillar-dynamische Untersuchungsergebnisse', Verlag Freies Geistesleben, Stuttgart, 1973).

- 'Die Signatur der Venus im Pflanzenreich' Stuttgart 1978.

- 'Die Signatur des Uranus im Pflanzenreich, Kapillar-dynamische Untersuchungsergebnisse', Stuttgart 1984.

Michel Gauquelin, 'The Cosmic Clocks, from Astrology to a Modern Science', 1969.

Walter Goldstein, 'The effect of planting dates and lunar positions on the yield of carrots' Biodynamics (US) July/August 2000, pp13-17.

Ursula Graf, PhD thesis, 'Darstellung vershiedener biologischer Landbaumethoden und Abklarung des Einflusses kosmischer Konstellationen auf das Pflanzenwachstum'. Zurich Technical College, Switzerland, 1977.

- and E.R. Keller, 'Zusammenhange zwischen kosmischen Konstellation und dem Ertrag landwirtschaftlicher Kulturplanzen...' Landw. Monath., 1979, 57, 10, pp325-36.

G. Grau, et al., 'Lunar phasing of the Thyroxine surge', Science 1981, 211, 607-9.

E. Graviou, 'Analogies between Rhythms in Plant Material in Atmospheric Pressure and solar-lunar periodicities' International Journal of Biometerology, 1978, 22.2.

- 'A Complex Rhythm problem: the Possibility of a lunar modulation of plant function,' Jnl. of Interdisc. Cycle Res. 1978, 9.

Maria Hachez, 'The Significance for Seed-Germination of the Passage of the Moon through the Constellations of the Zodiac', Anthroposophical Agricultural Foundation, (the precursor of Star and Furrow) November 1935, pp.283-289.

J.Herman and R.Golding, 'Sun, Weather and Climate' NASA 1978 (especially on geomagnetism and the solar cycle).

Hesiod 'Works and Days,' Loeb Classical Library, Heinemann 1977.

Hill, T. 'The Gardener's Labyrinth' 1577, reprinted OUP 1987.

Olga Holbeck, 'Variations in the form of Plant Buds,' Star and Furrow, Winter 1987 pp.17-20 (summarises two papers by Lawrence Edwards in Science Forum no. 5 (1985) and no. 7 (1987).

K. Holzknecht and E. Zurcher, 'Tree stems and tides – A new approach and elements of reflexion', Schweiz. Z. Forstwes. 2006, 157, 185-190.

Peter Huber, 'Uber den Nullpunkt der Babylonischen Ekliptik' Centaurus 1958, 5, pp.192-208.

Nicolas Joly, 'Le Vin du Ciel à la Terre' Paris 1997; Wine from Sky to Earth, 1999, Texas.

'What is Biodynamic wine? The quality, the taste, the terroir' E. Sussex, 2007.

Johannes Kepler, 'On Giving Astrology Sounder Foundations, A new short dissertation concerning cosmology, with a physical prognosis for the coming year 1602', translated in Judith Field's 'A Lutheran astrologer: Johannes Kepler', Archive for History of Exact Sciences, 1984,31, pp.190-268.

'Kepler's Astrology, The first Complete English translation of Tertius Interveniens and other astrological writings', trans. Ken Negus (at Princeton), 2008, MA.

'The Kimberton Hills Agricultural Calendar, a beginners guide to understanding the influence of cosmic rhythms in farming and gardening' (yearly), Ed. Sherry Wildfeuer, Kimberton Hills Publications, P.O.Box 155, Kimberton, PA, USA (renamed Stella Natura).

J.W. King, 'Solar Radiation Changes and the Weather', Nature, 245, 1973, 443-446.

- 'Weather and the earth's magnetic field' Nature 247, 1974, 131.

J.W.King, et al., 'Agriculture and Sunspots', Nature 252, 1974, 2-3.

Gunther Klein, 'Farewell to the Biological clock', Springer NY, 2007.

Dennis Klocek 'A Biodynamic Book of Moons' 1983 Biodynamic literature, Wyoming US (A poetic and meditative approach).

Koepf, Petterson and Schaumann, 'Biodynamic Agriculture', 1976, 3rd edition 1990; 4th Edition; 'Biologisch-Dynamische Landwirtschaft' (with Manon Haccius instead of Petterson), 1996.

Herbert Koepf, 'The Biodynamic Farm, Agriculture in the service of the Earth and Humanity' 1989, Anthroposophic Press, N.Y.

- 'Research in Biodynamic Agriculture: Methods and Results' in Biodynamic Farming and Gardening Association Inc., USA 1993.

M. Kokus, 'The 18.6-year Cycle in Droughts and Flood: A Review of the Climate Research of Robert G. Currie', Cycles, 39, August 1988, pp.189-191.

Eugene and Lili Kolisko, 'Agriculture of Tomorrow' Bournemouth, UK, Kolisko Archive Publications 1978 (first edition 1939), paperback 1982.

L. Kolisko, 'Der mond und das Planzenwachstum' 1927 in Wachsmuth, G. Ed., Gaia-Sophia Dornach, IV, 358-379. Same title, in Mitteilungen des Biologischen Institutes am Goetheanum Dornach: 1934, I, 19-21, II, 17-24; 1935, III, 17-19, IV, 3-14: English translation 'The Moon and Plant Growth,' 1936.

N. Kollerstrom: 'Zodiac Rhythms in Plant Growth (Potatoes)' Mercury Star Journal, Summer 1977, reprinted in Biodynamics, US, Winter 1993 'Testing the Lunar Calendar', pp.44-48.

- 'By the Light of the Silvery Moon, a look at Cosmic Influences in Agriculture', Farmer's Weekly, 22.12.1978.

- 'Moon Harmonics in Plant Growth,, Mercury Star Journal, Christmas 1978, London.

- 'Zodiac Rhythms in Plant Growth, Lettuce', Mercury Star Journal, Spring 1978.

- 'Plant Response to the Synodic Lunar Cycle: A Review,' Cycles, Bulletin of the Foundation for the Study of Cycles, Pittsburgh, PA, 1980, 31, 61-63.

- 'A Lunar Sidereal Rhythm in Crop Yield and its Phasing in the Zodiacal Circle', Correlation, (a journal of the Astrological Association) June 1981.

- 'Our Mathematical Moon, "It's a Question of Balance"' The Astrological Journal, Autumn 1983.

- 'Zodiac Rhythms in Plant Growth, III: Beans', Star and Furrow (journal of the Biodynamic Agriculture Association), Winter 1984.

- 'Wheat Germination and Lunar Phase: A Pilot Study', Correlation, May 1984.

- 'Star Rhythm in Crop Yield, A Computer Study' The Astrological Journal, Spring 1985.

- 'The Sidereal Zodiac, a Review with some Reflections', The Astrological Journal, reprinted in Journal of the Seasons of the New Zealand Astrological Society, Summer 1985.

- 'Testing the Lunar Calendar', Biodynamics, (Kimberton, USA) Winter 1993 44-48.

- 'The Star-Zodiac of Antiquity', Culture and Cosmos, Winter 1997 pp.5-22.

- and Gerhardt Staudenmaier, 'Mond-Trigon-Wirkung: eine statistiche Auswertung', Lebendige Erde, November 1998, pp.478-483. Translated in Harvests, the NZ Biodynamic journal, as 'Maria Thun's Trigons, What have Other Investigators found?' Winter 1999 pp.13-17.

- and G. Staudenmaier, 'Mond in Tierkreis: anders rechnen - andere Ergebnisse' Lebendige Erde Jan 2001, 48-9.

- and C. Power, 'Influence of the Lunar Cycle on Fertility on Two Thoroughbred Stud Farms' Equine Veterinary Journal, January 2000.

- 'Planting and Gardening by the Moon' (yearly) Foulsham.

- and G. Staudenmaier, 'Evidence for lunar-Sidereal Rhythms in Crop Yield: A Review' Biological Agriculture and Horticulture, An International Journal for Sustainable Production Systems, 2001, 19, 247-261. Reprinted in 'Harvests' the NZ Biodynamic journal, as 'Sowing by the Moon and Stars' 2003, 44, pp.8-15.

- Lunar Effect of Thoroughbred Mare Fertility: an analysis of 14 years of Data 1986-1999' in Biological Rhythm Research 2004 35, pp. 317-328.

- 'The fortnightly tree bud rhythms of Lawrence Edwards', Archetype, 11, September 2005.

- 'Star-Rhythm in Mistletoe Shape,' Elemente der Naturwissenshaft, 2011, 94, 119-124.

- 'A Lunar Cycle in Mare Fertility', Correlation, 2011, 27 (2), pp.38-47.

- and Mark Moodie, 'Potato sowings over a lunar eclipse', Star and Furrow, Summer 2012, pp.25-6.

P. Korringa, 'The Moon and Periodicity in Breeding of Marine animals', Ecological Monographs 1947, 17, pp.348-381.

Chris Knight, 'Blood Relations, Menstruation and the Origins of Culture', Yale 1991, Ch 10, Hunter's Moon.

J. Knight, 'Moon Up - Moon Down, the story of the Solunar Theory', Biodynamics (US), 1972.

T.M. Lai, 'Phosphorus and Potassium Uptake by Plants Relating to Moon Phases', Biodynamics, Summer 1976 (U.S. publication), pp.1-15.

Lee Lehman, 'The Book of Rulerships,' Whitford US, 1993.

M. Lethbridge, 'Relationship between thunderstorm frequency and lunar phase', Jnl. Geophysical Reserch, 1970, 75, 5153.

Llewellyn's 'Lunar Organic Gardener' 1993, St Paul Minnesota, UK distributor Foulsham.

Arnold Lieber, 'The Lunar Effect', Corgi 1979.

Rev. Franklin Loehr, 'The Power of Prayer on Plants', NY 1959.

D. Lopez, 'Ritmos de Periodo Largo en el Crecimiento de las Plantas,' Mem. Acad. Cien. Artes, Barcelona 1969, 396, pp.169-218.

J. Lücke, 'Untersuchungen uber den Einflus der Saatzeiten nach dem siderischen Kalender auf Ertrag und Qualitat von Hafer und Kartoffeln' 1982 (dissertation at Giessen).

Volkmar Lüst, 'Eininge Praxiserfahrungen mit den "Thun'schen Aussaattagen"' Lebendige Erde November 1984, pp.257-261.

T. Mannila, 'Lunar and planetary periodicity of failure years in Finland and in Sweden', J.Sci. Agric. Soc., Finland, 1980, 52, pp.393-402.

Gervase Markham, 'The Whole Art of Husbandry,' 1631 (translation of a continental work, originally published in the 1570s.)

Richard Markson, 'Considerations regarding solar and lunar modulation of geophysical parameters, atmospheric electricity, and thunderstorms,' Pure & Applied Geophysics, 1971, 84, 161.

- 'Geophysical influences in biological cycles', Jnl. Interdisciplinary Cycles Research, 1972, 3, 134. For a fuller account see Markson, 'Tree Potentials and external factors', as an appendix to Harold Saxton Burr, 'Blueprint for Immortality', Neville Spearman, London 1977.

K. Mather & J. Newall, 'Seed Germination and the Moon,' Jnl. Roy. Hort. Soc. 1941, 66, 358-66.

K. Mather, 'The Effect of Temperature and the Moon upon Seedling Growth,' Journal of the Royal Horticultural Society, 1942, 67, 264-70.

M.G. Maw, 'Periodicities in the Influences of Air Ions on the Growth of Garden Cress', Canadian Jnl. of Plant Science, 1967, 47, 499-505.

Walter and Abraham Menaker, 'Lunar Periodicity in Human Reproduction: a Likely Unit of Biological Time,' American Journal of Obstetrics and Gynacology, 1959, 77, pp.905-914.

Maurice Messengué, 'Of Men and Plants, The autobiography of the world's most famous plant healer', 1972 New York.

Neil Michelson, 'The American Sidereal Ephemeris 1976 to 2000', A.F.A. San Diego, California, 1981.

B. Millet and N. Moallem (2001) 'Growth rhythms and sap flow in Mandarin Orange tree', (in French) L'Arbre, 2000, Ed. M. Labrecque. 4th Int. Symposium on the Tree, Montreal, 97-103; cited in Zurcher and Holzknecht, 2006.

Dennis Millner and E.Smart, 'The Loom of Creation', 1975, 1977.

El Hassania Mohssine et al., 'Lunar Phase Influence on the Glycemia of Worker Honeybees', Chronobiologica 1990, 17, 201-7.

R.S. Nishioka et al., 'Effect of lunar-phased release on Adult Recovery of Coho Salmon ... in California', Aquaculture 82 (1989) pp.355-365.

M.G. Oehmke, 'Lunar Periodicity in Flight Activity of Honey Bees' Journal. of Interdisciplinary Cycle Research, 1973, 4, 319-335.

Jane Panzer, 'Lunar Correlated Variations in Water Uptake and Germination in three species of seeds' PhD Thesis, Tulane University, 1976.

Johanna Paungger and Thomas Poppe, 'Moon time, the art of harmony with nature and lunar cycles', 1993 Germany, 1995 English trans.

Alex Podolinsky, 'Biodynamic Agriculture Introductory Lectures', Vol 1, Fourth Edn 1990, Victoria Australia.

Pliny the Younger, 'History of Nature', (Historia Naturalis) Volume 18, section 75.

U.J. Pittman, 'Magnetism and Plant Growth, III, Effect on germination and early growth of corn and beans', Canadian Journal of Plant Science, 1965, 45, 549-555.

Robert Powell, 'Lunar Calendar for Farmers and Gardeners', Mercury Star Journal, Summer 1977.

- and Peter Treadgold, 'The Sidereal Zodiac', Anthroposophical Publications, London, 1979, reprinted by American Federation of Astrologers, 1985.

- 'The Zodiac: A Historical Survey' (16 pages) 1985 Astro Computing Services, San Diego California.

- and Kavin Dann, 'The Astrological Revolution', Lindisfarne Books, 2010, Appendix 4 'Planting by the Moon.'

A.Presman, 'Electromagnetic Fields and Life,' Plenum Press, New York.

M. Rampino et al., Eds., 'Climate: History, Periodicity ands Predictability,' NY 1987; especially Ch. 22 by R.G. Currie, 'Examples and Implications of 18.6 and 11-yr Terms in World Weather Records.'

Louise Riotte, 'Planetary Planting', 1975, 1982 paperback, Astro-computing services, US.

R.M. Ross, 'Reproductive Behaviour of the Anemonefish,' Copelia 1 (1978), pp.103-107.

M. Rossignol et al., 'Lunar cycle and nuclear DNA variations in potato callus or root meristem', Geo-cosmic Relations, Ed. Tomassen, Wageningen, 1989, 116-126.

Harry Rounds, 'A Semi-lunar periodicity of neurotransmitter-like substances from plants' Physiologica Plantarium, 1982, 54, 495-9.

F. Sattler & E. Wistinghausen, 'Biodynamic Farming Practice', CUP, 1989 (1985 in German).

W. Schaumann and H. Spiess, 'Mitteilung in Zisammenhang mit dem Bericht von V. Lust', Lebendige Erde, p.261-262.

Joachim Schultz, 'Movement and Rhythm of the Stars' 1963, Trans. 1986 Floris Books.

Theodor Schwenk, 'Sensitive Chaos' 1962, translated 1965.

Jochen Schuchow, John Wilkes & Ian Trousell, 'Energising Water, Flowform Technology and the Power of Nature', Sophia Books, 2010.

M.T. Sherill and D. Middaugh, 'Spawning Periodicity of the Inland Silverside .. in the Laboratory: Relationship to Lunar Cycles', Copeia, 2 (1993), pp.522-528.

Richard Thornton Smith, 'Cosmos, Earth and Nutrition, The Biodynamic Approach to Agriculture', Sophia Books, 2009.

John Soper, 'Biodynamic Gardening', 1983, Biodynamic Agricultural Association, Stourbridge,

Hartmut Spiess, 'Zur Frage der Wirksamkeit kosmischer Rhythmen und Konstellationen', Lebendige Erde, 1987, 6, pp.305-315.

- 'Konventionelle und Biologische-Dynamische Verfahren zur Steigerung der Bodenfruchtbarkeit'. Dissertation, 1978, Giessen.

- 'Chronobiological Investigations of Crops Grown under Biodynamic Management. I. Experiments with Seeding Dates to Ascertain the Effects of Lunar Rhythms on the Growth of Winter Rye' Biological Agriculture and Horticulture, 1990,7, pp.165-178; II. '..on the growth of Little Radish', Ibid, pp.179-189.

- 'Haben lunare Rhythmen Bedeutung fur den Okologischen Landbau?' 1993 pp397-403.

- (1994) 'Chronobiologische Untersuchungen mit besonderer Berucksichtigung lunarer Rhythmen im biologische-dynamischen Pflanzenbau', Darmstadt; 2 vols, I 'Band 3' of 258 pp. , and II 'Band 4 - beschreibung der einzelergebnisse' of 319 pp., the data.

- 'Lunar Rhythms and Plants', Biodynamics May/June 2000, 19-22.

- 'Mondrhythmen wirken – Trigoneinflusse nicht gefunden' Lebendige Erde, 2001, 1, 18-21.

Rudolf Steiner, 'Agriculture' (lectures), 1927 (in manuscript) 1958 published in English, reprinted 1984, Biodynamic Agriculture Association.

H.L. Stolov and A. Cameron, 'Variations of Geomagnetic Activity with lunar phase,' Journal of Geophysical Research, 1964, 69, 4975-4982.

E. Tavenner, 'The Roman Farmer and the Moon,' Transactions and Proceedings of the American Philological Association, 1918, 49, 67-82.

Jack Temple, 'Gardening Without Chemicals', 1986.

- 'Checking the value of Planting by the Zodiac', Here's Health, November 1982 pp.144-5.

B. Timmins, 'Planting by the Moon', Aries Press, Chicago, 1937.

Louis Thompson, 'The Moon may affect Agricultural ups and downs', Farm Journal, August 1987, 24.

- 'The 18.6-year Lunar Cycle: its Possible Relation to Agriculture' Cycles, (Pittsburgh, US) March 1989 65-9.

- 'Sunspots and Lunar Cycles: Their possible relation to Weather cycles' Cycles, September 1989, 26- 5-268.

Peter Tompkins and Christopher Bird, 'The Secret Life of Plants', 1973.

- 'Secrets of the Soil' 1991, Penguin Books.

Maria Thun, 'Nine Years Observation of Cosmic Influences on annual Plants' (Trans from Lebendige Erde, Jan/Feb 1963, Star and Furrow 22, Spring 1964).

- 'Aussaattage' (yearly since 1963, now translated into 22 languages); English edition is entitled, 'Working with the Stars' and appears in January (NB, this title is by no means a translation of the German).

- 'Cosmic Working of Soil and Plant of Sidereal Moon-Rhythms', Mercury Star journal, Summer 1978, pp. 42-51 (Translated from the Sternkalendar, Dornach 1973).

- 'Work on the Land and the constellations', Lanthorne Press 1977, East Grinstead, Sussex; 2nd Edn 1991, 70 pp (3rd Edition, Hinweise aus der Konstellationsforschung 1994, 210 pp).

- and Hans Heinze, 'Mondrhythmen im siderischen Umlauf und Pflanzenwachstum', published by Forschungsring fur biologisch-dynamische Wirtschaftsweise, 6100 Darmstadt-Land 3, 1979.

- 'Milch und Milchverarbeitung', 1985, 1991 Stuttgart.

- 'Erfahrungen fur den Garten', 1994 Stuttgart; English translation, 'Gardening for life - the Biodynamic way', Hawthorne Press UK, 1999.

- Gardening for Life: the Biodynamic Way 2002

- 'When Wine Tastes Best' (with Matthias Thun) 2009 Floris books.

- C. Timmins, 'Planting by the Moon', Aries Press, Chicago, 1939 (I have found no copy in the UK).

R.G. Vines, 'Possible Relationships between Rainfall, Crop Yields and the sunspot Cycle', Jnl. Australian Inst. of Agricultural Science, March 1977, 3-13.

H. Vlasinova et al., 'The Mitotic Activity of Norway Spruce polyembyonic culture oscillates during the synodic lunar cycle'. Biologia Plantarum 2003, 47, 475-6.

Kristiina Vogt et al., 'Indigenous Knowledge Informing Management of Tropical forests: the Link between Rhythms in plant secondary chemistry and lunar cycles', Ambio, Sept 2002 31, 485-90.

Rudolf Vollman, 'The Menstrual Cycle', 1977, Volume 7 in the series, Major Problems in Obstetrics and Gynecology, Ed. Friedman.

Elisabeth Vreede, 'The Ascending paths of the planets at their times of special influence' (translated from the Calendar of 1935/6) Anthroposophical Agricultural Foundation, Notes & Correspondence, June 1936.

Guenther Wachsmuth, 'The Etheric Formative Forces in Cosmos, Earth and Man,' London Anthroposophic Press 1932.

- 'Erde und Kosmos' 1952, Dornach.

Christopher Walker Ed., 'Astronomy Before the Telescope' 1996, Ch.3, Astronomy & Astrology in Mesapotamia (star-zodiac).

Sherry Wildfeuer, 'Stella Natura, the Kimberton Hills Agricultural Calendar' 19th Edn, 1986, Kimberton USA.

F. Wylie, 'Tides and the Pull of the Moon', 1980 Berkeley Books.

Ernst Zurcher, 'Rhythmicities in the Germination and Initial Growth of a Tropical Forest Tree Species', (in French) Journal Forestier Suisse, 1992, 143, 951-966.

- 'Le rhthme synodique lunaire et la croissance végétale: Bilan d'essais sur les rhythmicities dans la germination et la croissance initiale d'essences foresti_res tropicales' L'Arbre, Biologie et Développement, Troisieme Colloque Internationale Monpellier 1995, Ed. C. Edelin, 150-164.

- et al., 'Tree Stem diameters fluctuate with tide', Nature 1998, 392, 665-6.

- 'Lunar-Related traditions in Forestry and Phenomena in Tree Biology' (in German), Journal Forestier Suisse, 2000, 151 417-424.

- and Daniel Mandallaz,"Lunar Synodic Rhythm and Wood Properties: Traditions and Reality - Experimental Results on Norway Spruce', L'Arbre 2000, 4th International Symposium on the Tree, Montral, 2001, 244-250.

- 'Lunar Rhythms in forestry Traditions - Lunar-Corelated Phenomena in tree biology and wood properties,' Earth, Moon and Planets, an International Journal of Solar System Science, Kluwer Academic, Holland 2001, 85, 463-478.

- 'Cosmic trees and traditional knowledge of lunar rhythms,' in Moving Worldviews - Reshaping sciences, policies and practices for endogenous sustainable development, Bertus Haverkort and Coen Reijntjes (eds), Leusden 2006 and K Holznecht, 'Tree stems and tides – A new approach and elements of reflexion', Schweiz. Z. Forstwes. 157 (2006), 6, 185-190.

- 'Plants and Moon,' a chapter in 'Aux origines des Plantes', Francis Hallé (ed.), Paris 2008.

- Lunar-synodic variation in the germination of European spruce (Picea abies) seeds: a previous trial re-evaluated (in preparation).

- et al., 'Looking for differences in wood properties as a function of the felling date: lunar phase-correlated variations in the drying behaviour of Norway Spruce and Sweet Chestnut', Trees (2010) 24:31–41

- 'Plants and the Moon – traditions and phenomena', HerbalGram, journal of the American Botanical Council, Volume 8, Number 4, April 2011 (online).

Index

Abele, Ulf, 71, 73, 220
Abrami, Giorgio, 49
Addey, John, 191
Agriculture of Tomorrow, Kolisko, 54, 92
Aldebaran, 139
Algol, 24
American Federation of Astrologers, 103
anthroposophical, 129
Anthroposophical, 220
Anthroposophists, 76
Apollo 11 landing, 216
Aristotle, 9
ascending Moon, 40
Avebury, 37
ayanamsa, 228
Backster, Cleve, 214
Bacon, Frances, 53
Barger, Adele, 150
bee activity, 204
Bees, 52
Best, Simon, 5, 176
Biodynamic system, 20
Bishop radish data, 231
Bishop, Colin, 8, 112, 198
Book of Revelations, 128
British Museum, 132
broad beans, 116
Brown, Frank, 45, 197
Bunning, 43
Burns, John T., 225

Burr, Harold, 51
Campion, Nicholas, 20
Canadian lynx, 208
Chaldeans, 130
chronobiology, 43, 181
Culpeper's Herbal, 160
Dariotus, 149, 160
De Fundaments Astrologiae Certioribus, 12
Demeter, 22
Dewey, John, 207
DNA, of plants, 13, 52
Dubrov, Alexandre, 200
Ecclesiastes, 3
eclipse seasons, 176
ecliptic, 33, 36
Edwards, Lawrence, 57, 164, 167, 225, 229
eels and GMF, 184
Eliott, Jean, 156
Enuma Anu Enlil, 192
Equine Fertility Unit, Newmarket, 175
Farmers' Weekly, 194
fertility, of mares, 174
fishing almanacs, 182
formative force, 219
four elements, 67
four ethers, 228
Fyfe, Agnes, 14, 161
Gaia, 210
Gauquelin, Michel, 155

geomagnetotropism, 201
gestation period, 15
GMF, 183, 199
goats' milk, 142
Goethe, 99
Goethean, 143
Graf, Ursula, 72, 221
Graviou, Elaine, 46
Great Bull of Heaven, 131
Hamaker-Zondag, Karen, 190
Harmony by HRH Charles, 215
Harvard University, 214
Haushka, Rudolf, 101
Heinze, Hans, 69, 75, 78
Herodotus, 130
Hiscia Klinik, 90, 225, 230
horse fertility, 205
Ides of March, 28
International Astronomical Union, 138
International Wine Fair, 146
John Innes Horticultural foundation, 54
Joly, Nicholas, 157, 159
Kepler, 39
Kepler, Johannes, 11
Kervran, Louis, 101
Kimberton Hills calendar, 95, 140
Kolisko, 53, 205
Kolisko, Lili, 14
Korringa, 18
Lai, T.M., 50
lambda-function, 225
Lebendige Erde, 83
Lehman, Lee, 152, 162
Leo constellation, 190
lettuce, 123
lettuce seed, 118

Lllewellyn almanac, 128
Lücke, 74
lunar day, 46
lynx-pelt abundance, 207
mad cow disease, 24, 213
Maria Thun, 22
Markson, Ralph, 51
Mars, 216
Mather, K., 55, 217
Maunder Minimum, 195
Maw, M.G., 55
Merchant, Caroline, 215
Mesapotamia, 131
Michelson, Neil, 91
mistletoe berries, 227
Moodie, Mark, 193
Moon Sign Book, 89
Moonrise, 199
Mother Nature, 213
moving average, 175
Muntz, Reg, 8, 109, 115
Newmarket stud farms, 16
Newton, Isaac, 30, 215
Nietzsche, 41
nodal cycle, 34
Norfolk Punch, 160
Oehmke, 52, 204
Ophiuchus, 139
organic gardeners, 3
Panzer, Jane, 46
perigee, 224
Pisces, 141
Planting by the Moon, 91, 148
Platt, Paul, 140
Pliny, 136, 188
Pliny the Elder, 10
Plutarch, 48
Podolinsky, Alex, 186
Ptolemy, Claudius, 17, 64, 69
Racehorses, 170

Renaissance vision, 149
Riotte, Louis, 105, 106
Royal Astronomical Society, 183
Rudolf Steiner, 26
Rulni calendar, 89, 91
Rulni, Franz, 192
salmon, 182
Salmon fisheries, 181
Schmidt, Georg, 154
Schwenck, Theodore, 192
sea-horses, 182
Sellar, Wanda, 154
Shakespeare, 30, 192
Sidereal month, 28, 39
Sidereal zodiac, 137
Silver Axioms, 180
Soil Association, 22
solar radiation, 199
solar-lunar wave-functions, 202
Spica, 133, 159
Spiess, Hartmut, 79, 81, 93, 107, 221
star-zodiac, 66
Staudenmaier, Dr, 83
Steiner, Rudolf, 151, 212, 226
synodos, 30

Tablet of Hermes, 209
Taurus the Bull, 135
Temple, Jack, 94
Termites nests, 184
Tetrabiblos, 18
Thales of Miletus, 10
The Secret Life of Plants, 100
third harmonic sinewave, 230
Thun calendar, 7, 97
Thun theory, 83, 119
Thun, Maria, 65, 142, 190, 211
tomato growth, 187
tree cell mitosis, 203
trigons, 133
tropic of Cancer, 38
Vernal Equinox, 135, 182
Vettius Valens, 146
Vreede, Elizabeth, 138
Wachsmuth, 219
wheat germination, 56
When Wine Tastes Best,, 144
Wildfeuer, Sherry, 95
wines, German, 197
Zodiac signs, 65
Zürcher, 59, 61
Zurcher, Ernst, 42, 58, 89, 210, 219

www.ingramcontent.com/pod-product-compliance
Lightning Source LLC
Chambersburg PA
CBHW071228080526
44587CB00013BA/1536